TC 3-21.60

VISUAL SIGNALS

MARCH 2017

DISTRIBUTION RESTRICTION: Approved for public release; distribution is unlimited.

*This publication supersedes FM 21-60, 30 September 1987.

Headquarters, Department of the Army

This publication is available at the Army Publishing
Directorate site (http://www.apd.army.mil),
and the Central Army Registry site
(https://atiam.train.army.mil/catalog/dashboard)

*TC 3-21.60 (FM 21-60)

*Training Circular
No. 3-21.60 (FM 21-60)

Headquarters
Department of the Army
Washington, DC, 17 March 2017

Visual Signals

Contents

Chapter 1	HAND-AND-ARM SIGNALS FOR DISMOUNT OPERATIONS	1-1
	General	1-1
	Signals for Dismounted Formations	1-1
	Patrolling Hand and Arm Signals	1-7
	Signals for Crew-Served Weapons	1-18
Chapter 2	HAND-AND-ARM SIGNALS FOR GROUND VEHICLES	2-1
	General	2-1
	Signals for Mechanized Movement Techniques	2-1
	Signals to Control Vehicle Drivers and/or Crews	2-11
	Traffic Control	2-19
	Convoy Control	2-21
	Flag Signals	2-24
	Firing Range Flag Signals	2-27
Chapter 3	HAND AND ARM SIGNALS FOR AIRCRAFT	3-1
	General	3-1
	Common Hand and Arm Ground Signals	3-1
	Ground-to-Air Panel System	3-12
	Special Panel Signals	3-14
	Ground-to-Air Emergency Signals and Codes	3-16
	Signaling With Mirrors and Strobes	3-18
Chapter 4	PYROTECHNICS	4-1
	General	4-1
	Description	4-1
	Handheld Signals	4-1
	GLOSSARY	Glossary-1
	REFERENCES	References-1
	INDEX	Index-1

Figures

Figure 1-1. Disperse ... 1-1
Figure 1-2. Assemble or rally ... 1-2

Distribution Restriction: Approved for public release; distribution is unlimited.

*This publication supersedes FM 21-60, 30 September 1987.

i

Contents

Figure 1-3. Join me, follow me, or come forward 1-2
Figure 1-4. Increase speed, double time, or rush 1-3
Figure 1-5. Wedge formation 1-3
Figure 1-6. Vee formation 1-4
Figure 1-7. Line formation 1-4
Figure 1-8. Echelon left 1-5
Figure 1-9. Echelon right 1-5
Figure 1-10. Staggered column formation 1-6
Figure 1-11. Column formation 1-6
Figure 1-12. Herringbone formation 1-7
Figure 1-13. Contact left 1-7
Figure 1-14. Contact right 1-8
Figure 1-15. Air attack 1-8
Figure 1-16. Chemical, biological, radiological, and nuclear attack 1-9
Figure 1-17. Fix bayonets 1-9
Figure 1-18. Enemy in sight 1-10
Figure 1-19. Quick time 1-10
Figure 1-20. Take cover 1-11
Figure 1-21. Map check 1-11
Figure 1-22. Halt 1-12
Figure 1-23. Take a knee 1-12
Figure 1-24. Move to the prone 1-13
Figure 1-25. Pace count 1-13
Figure 1-26. Radiotelephone operator forward 1-14
Figure 1-27. Head count 1-14
Figure 1-28. Danger area 1-15
Figure 1-29. Freeze 1-15
Figure 1-30. Move platoon leader to the front 1-16
Figure 1-31. Move the platoon sergeant to the front 1-16
Figure 1-32. Move squad leader forward 1-17
Figure 1-33. Number signals 1-17
Figure 1-34. Stop, look, listen, smell 1-18
Figure 1-35. Message acknowledged 1-18
Figure 1-36. Commence firing or change rate of fire 1-19
Figure 1-37. Change direction or elevation 1-19
Figure 1-38. Move over or shift fire 1-20
Figure 1-39. Cease firing 1-20
Figure 1-40. Out of action 1-21
Figure 2-1. Wedge formation 2-1
Figure 2-2. Vee formation 2-2
Figure 2-3. Line 2-2

Figure 2-4. Coil	2-3
Figure 2-5. Echelon left	2-3
Figure 2-6. Echelon right	2-4
Figure 2-7. Staggered column formation	2-4
Figure 2-8. Column formation	2-5
Figure 2-9. Herringbone formation	2-5
Figure 2-10. Traveling	2-6
Figure 2-11. Traveling overwatch	2-6
Figure 2-12. Bounding overwatch	2-7
Figure 2-13. Fire	2-7
Figure 2-14. Move to the left	2-8
Figure 2-15. Move to the right	2-8
Figure 2-16. Advance, move out or "follow me"	2-9
Figure 2-17. Dismount	2-9
Figure 2-18. Stop	2-10
Figure 2-19. Button up or unbutton	2-10
Figure 2-20. Open up	2-11
Figure 2-21. Close up	2-11
Figure 2-22. Attention	2-12
Figure 2-23. I am ready or ready to move or are you ready?	2-12
Figure 2-24. Mount	2-13
Figure 2-25. Disregard previous command or remain in place	2-13
Figure 2-26. I do not understand	2-14
Figure 2-27. Start engine or prepare to move	2-14
Figure 2-28. Halt or stop	2-15
Figure 2-29. Increase speed	2-15
Figure 2-30. Right or left turn	2-16
Figure 2-31. Slow down	2-16
Figure 2-32. Move forward	2-17
Figure 2-33. Move in reverse (for stationary vehicles)	2-17
Figure 2-34. Close distance between vehicles and stop	2-18
Figure 2-35. Stop engines	2-18
Figure 2-36. Neutral steer (track vehicles)	2-19
Figure 2-37. Left and right traffic stop	2-19
Figure 2-38. Front traffic stop	2-20
Figure 2-39. Rear traffic stop	2-20
Figure 2-40. Traffic from right, GO	2-21
Figure 2-41. Traffic from left, GO	2-21
Figure 2-42. Open up or increase the distance between vehicles	2-22
Figure 2-43. Close up or decrease the distance between vehicles	2-22
Figure 2-44. Pass and keep going	2-23

Contents

Figure 2-45. Move in reverse 2-23
Figure 2-46. Mount 2-24
Figure 2-47. Dismount 2-25
Figure 2-48. Dismount and assault 2-25
Figure 2-49. Assemble or close 2-26
Figure 2-50. Move out 2-26
Figure 2-51. Chemical, biological, radiological, and nuclear hazard present 2-27
Figure 2-52. All weapons clear (guns elevated) 2-27
Figure 2-53. Conducting live fire or "hot gun" 2-28
Figure 2-54. Conducting prepare-to-fire or non-firing exercises (Ammunition is uploaded and the system is on safe.) 2-28
Figure 2-55. Malfunction—weapons clear 2-29
Figure 2-56. Malfunction—weapons loaded 2-29
Figure 3-1. Helicopters (rotary wing) 3-1
Figure 3-2. Fixed-wing aircraft 3-2
Figure 3-3. Assume guidance 3-2
Figure 3-4. Cut engine(s) or stop rotor(s) 3-3
Figure 3-5. Emergency signal 3-3
Figure 3-6. Proceed right to next signalman 3-4
Figure 3-7. Proceed left to next signalman 3-4
Figure 3-8. Depart 3-5
Figure 3-9. Go around, do not land 3-5
Figure 3-10. Land 3-6
Figure 3-11. Stop 3-6
Figure 3-12. Spot turn 3-7
Figure 3-13. Move right 3-7
Figure 3-14. Move left 3-8
Figure 3-15. Move ahead 3-8
Figure 3-16. Move rearward 3-9
Figure 3-17. Load has not been released 3-9
Figure 3-18. Hookup complete 3-10
Figure 3-19. Release 3-10
Figure 3-20. Move downward 3-11
Figure 3-21. Move upward 3-11
Figure 3-22. Hover 3-12
Figure 3-23. NATO standard panel code figures for numbers 3-13
Figure 3-24. Aircraft acknowledgement 3-13
Figure 3-25. Wind direction, wind-T 3-14
Figure 3-26. Pick up message here (wind in direction indicated) 3-15
Figure 3-27. Dropped message not received 3-15
Figure 3-28. Enemy aircraft in your vicinity 3-16

Figure 3-29. Direction indicator 3-16
Figure 3-30. Emergency signals 3-17
Figure 3-31. Emergency codes 3-18
Figure 3-32. How to use a signal mirror 3-19
Figure 4-1. Star clusters 4-2
Figure 4-2. Single star 4-2

Tables

This section contains no entries.

This page intentionally left blank.

Preface

The purpose of Training Circular (TC) 3-21.60 is to standardize visual signals and to serve as a training reference. It is a guide. It does not cover all visual signals used in the Army, only those that are commonly used. Signals used with equipment or during operations are in manuals that relate to such operations.

Efficient combat operations depend on clear, accurate, and secure communication among ground units, Army aviation, and supporting Air Force elements. Control and coordination are achieved by the most rapid means of communication available between Soldiers and units. When electrical and/or digital means of communication are inadequate, or not available, a station-to-station system of visual communication is an alternate means for transmitting orders, information, or requests for aid or support.

Through the use of hand-and-arm signals, flags, pyrotechnics, and other visual aids, messages may be transmitted. Although many of these signals are widely used, incorporated into unit communications-electronics operating instructions and standing operating procedures, Army-wide standardization will increase their effectiveness.

The principal audience for TC 3-22.240 is all members of the profession of arms. Commanders and staffs of Army headquarters serving as joint task force or multinational headquarters should also refer to applicable joint or multinational doctrine concerning the range of military operations and joint or multinational forces. Trainers and educators throughout the Army also will use this publication.

Commanders, staffs, and subordinates ensure their decisions and actions follow applicable United States (U.S.), international, and, in some cases, host-nation laws and regulations. Commanders at all levels ensure their Soldiers operate according to the law of war and rules of engagement. (Refer to Field Manual (FM) 27-10 for more information).

Terms for which TC 3-21.60 is the proponent publication (the authority) are italicized in the text and are marked with an asterisk (*) in the glossary. Terms and definitions for which TC 3-21.60 is the proponent publication are boldfaced in the text. For other definitions shown in the text, the term is italicized and the number of the proponent publication follows the definition.

This manual applies to the Active Army, the Army National Guard (ARNG)/National Guard of the United States (ARNGUS), and the U.S. Army Reserve (USAR) unless otherwise stated.

Unless otherwise stated, whenever the masculine gender is used, both men and women are included.

The proponent for this publication is the U.S. Army Training and Doctrine Command (TRADOC). The preparing agency is the U.S. Army Maneuver Center of Excellence (MCoE). Send comments and recommendations by any means, U.S. mail, e-mail, fax, or telephone, using the format of DA Form 2028, *Recommended Changes to Publications and Blank Forms*. Point of contact information is as follows.

E-mail:	usarmy.benning.mcoe.mbx.doctrine@mail.mil
Phone:	COM 706-545-7114 or DSN 835-7114
Fax:	COM 706-545-8511 or DSN 835-8511
U.S. Mail:	Commander, MCoE
	Directorate of Training and Doctrine (DOTD)
	Doctrine and Collective Training Division
	ATTN: ATZB-TDD
	Fort Benning, GA 31905-5410

This page intentionally left blank.

Introduction

Visual signals are any means of communication that require sight and can be used to transmit prearranged messages rapidly over short distances. This includes the devices and means used for the recognition and identification of friendly forces.

The most common types of visual signals are hand-and-arm, flag, pyrotechnic, and ground-to-air signals. However, Soldiers are not limited to the types of signals discussed and may use what is available. Chemical light sticks, flashlights, and other items can be used provided their use is standardized within a unit and understood by Soldiers and units working in the area.

> **Signal Similarity**
>
> Visual signals are generally contextual in nature. For example, the hand-and-arm signal for "take cover" and "slow down" are similar in their perspective movements, however the situation in which each is given is completely different. Hand-and-arm signals are not exclusive to dismounted or mounted formations as long as the message is sent and received accurately. The only limit is the Soldier's initiative and creativity.

Visual signals have certain limitations—
- The range and reliability of visual communications are significantly reduced during periods of poor visibility and when terrain restricts observation.
- They may be misunderstood.
- They are vulnerable to enemy interception and may be used for deception purposes.

> **Weapon Control**
>
> Unless otherwise depicted, the Soldier's individual weapon has been removed from the figure for clarity. Soldiers must maintain positive control of their weapon at all times. If it is not feasible to perform the proper hand-and-arm signal while maintaining positive control, an alternate signal should be used with the non-firing hand. In the event that the Soldier must use both hands to perform a signal, the weapon will be moved to the hang or sling arms position.

This page intentionally left blank.

Chapter 1
Hand-and-Arm Signals for Dismount Operations

GENERAL

1-1. Signals illustrated with a single arrowhead indicate that the signal is not continuously repeated; however, it may be repeated at intervals until acknowledged or the desired action is executed. Signals illustrated with double arrowheads are repeated continuously until acknowledged or the desired action is taken. Signals are illustrated as normally seen by the viewer. Some signals are illustrated in oblique, right angle, or overhead views for clarity.

SIGNALS FOR DISMOUNTED FORMATIONS

1-2. Leaders of dismounted units use hand and arm signals to control the movement of individuals, teams, and squads. (See figures 1-1 through 1-12 on pages 1-1 through 1-7.)

1-3. To signal "disperse," extend either arm horizontally from the shoulder: wave the arm repeatedly to the front and to the side in a sweeping motion with the palm toward ground. (See figure 1-1.)

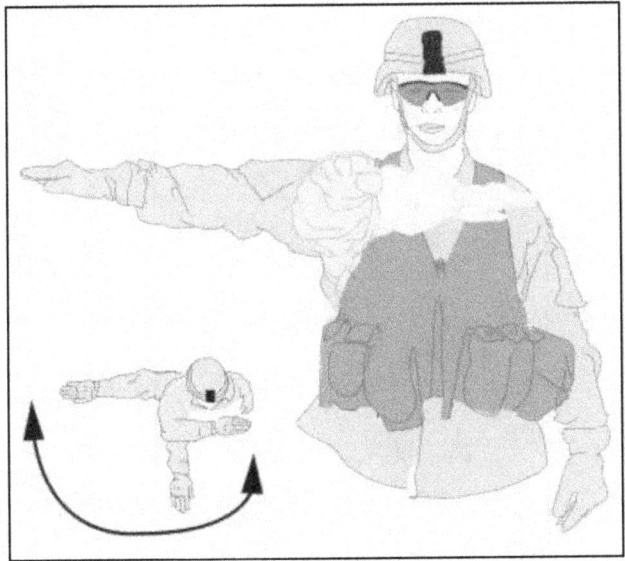

Figure 1-1. Disperse

Chapter 1

1-4. To signal "assemble" or "rally," raise the arm vertically overhead, palm to the front, and wave in large, horizontal circles. (See figure 1-2.)

Note. Signal is normally followed by the signaler pointing to the assembly or rally site.

Figure 1-2. Assemble or rally

1-5. To signal "join me," "follow me," or "come forward," point toward the person(s) or unit(s); beckon by holding the arm horizontally to the front, palm up, and motioning toward the body. (See figure 1-3.)

Figure 1-3. Join me, follow me, or come forward

Hand and Arm Signals for Dismounted Operations

1-6. To signal "increase speed," "double time" or "rush," raise the fist to the shoulder; thrust the fist upward to the full extent of the arm and back to the shoulder level; do this several times rapidly. (See figure 1-4.)

Figure 1-4. Increase speed, double time, or rush

1-7. To signal "wedge formation," extend the arms downward and to the sides at a 45-degree angle below the horizontal, palms to the front. Alternately, use the non-firing hand and point the index finger and pinky finger to the ground. All other fingers will be curled. (See figure 1-5.)

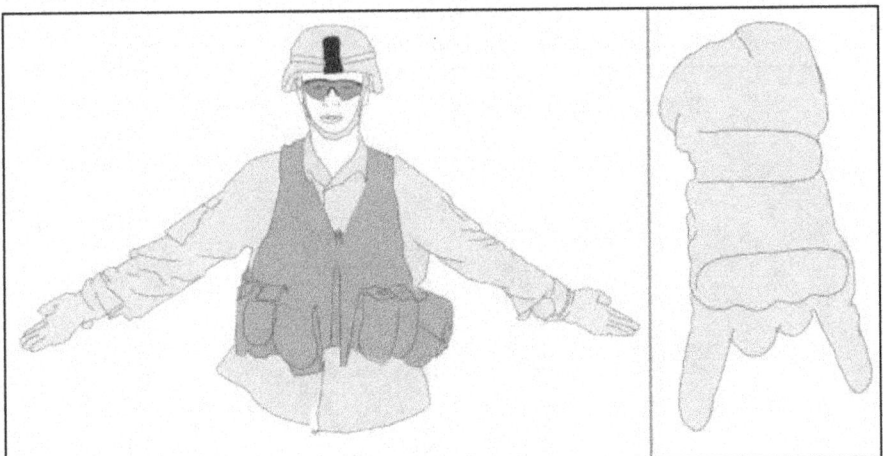

Figure 1-5. Wedge formation

Chapter 1

1-8. To signal "vee formation," raise the arms and extend them 45 degrees above the horizontal. Alternately, use the non-firing hand and point both the index and pinky finger up. All other fingers will be curled. (See figure 1-6.)

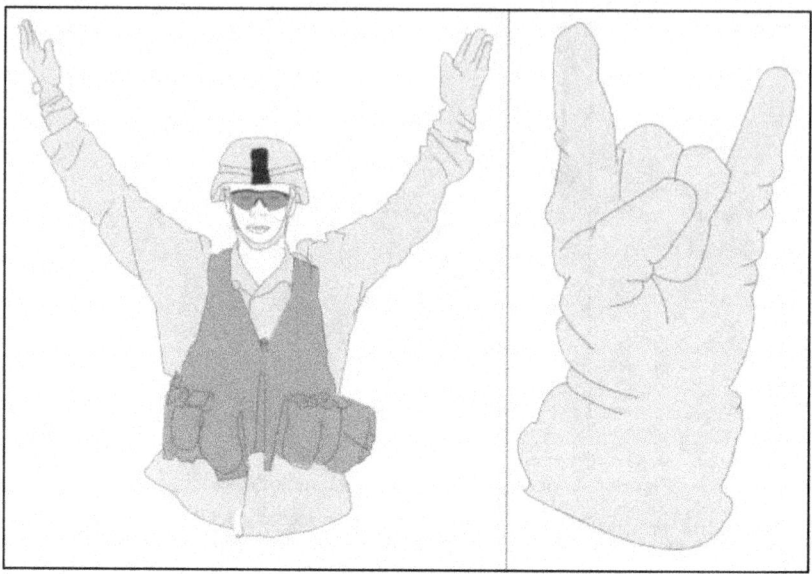

Figure 1-6. Vee formation

1-9. To signal "line formation," extend the arms parallel to the ground. (See figure 1-7.)

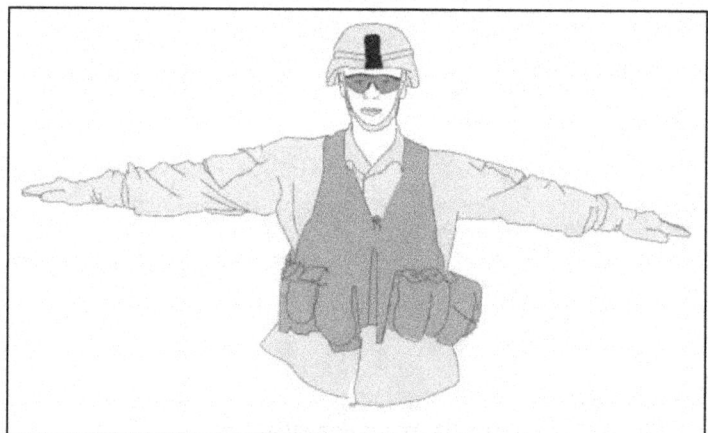

Figure 1-7. Line formation

1-10. To signal "echelon left," extend the right arm and raise it 45 degrees above the shoulder. Extend the left arm 45 degrees below the horizontal and point toward the ground. (See figure 1-8.)

Figure 1-8. Echelon left

1-11. To signal "echelon right," extend the left arm and raise it 45 degrees above the shoulder. Extend the right arm 45 degrees below the horizontal and point toward the ground. (See figure 1-9.)

Figure 1-9. Echelon right

1-12. To signal "staggered column formation," extend the arms so that upper arms are parallel to the ground and the forearms are perpendicular. Raise the arms so they fully extended above the head. Repeat. (See figure 1-10.)

Figure 1-10. Staggered column formation

1-13. To signal "column formation," raise and extend the arm overhead. Move it to the right and left. Continue until the formation is executed. Alternately, to signal form a file or column, move the non-firing hand to touch the rim of the headgear directly in front of the face. Fingers will be extended and joined. (See figure 1-11.)

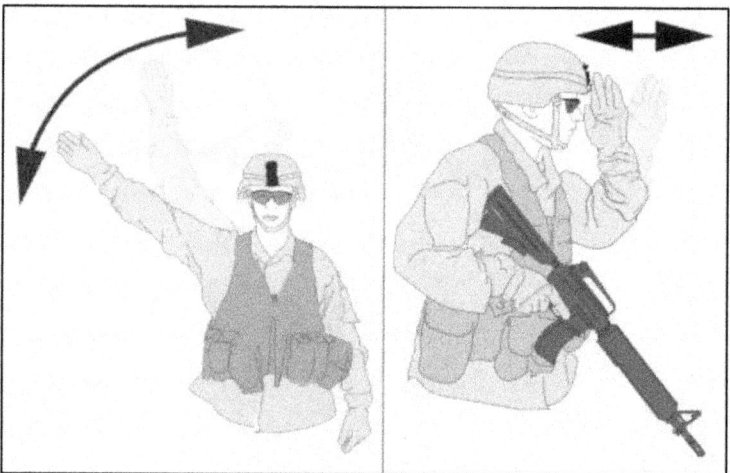

Figure 1-11. Column formation

1-14. To signal "herringbone formation," extend the arms parallel to the ground. Bend the arms until the forearms are perpendicular. Repeat. (See figure 1-12.)

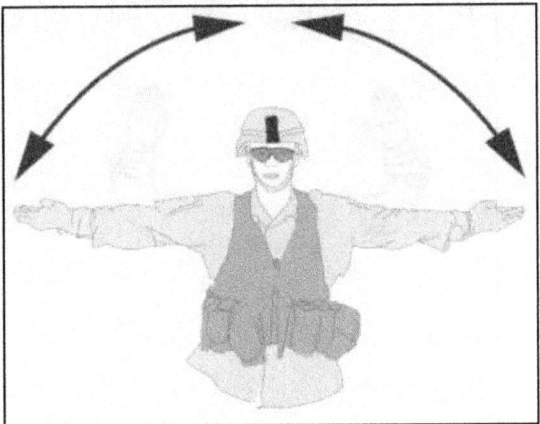

Figure 1-12. Herringbone formation

PATROLLING HAND AND ARM SIGNALS

1-15. Patrolling is conducted by many type units. Infantry units patrol in order to conduct combat operations. Other units patrol for reconnaissance and security. Successful patrols require clearly understood communication signals among patrol members (see figures 1-13 through 1-35 on page 1-7 through page 1-18.).

1-16. To signal "contact left," extend the left arm parallel to the ground. Bend the arm until the forearm is perpendicular. Repeat. (See figure 1-13.)

Figure 1-13. Contact left

1-17. To signal "contact right," extend the right arm parallel to the ground. Bend the arm until the forearm is perpendicular. Repeat. (See figure 1-14.)

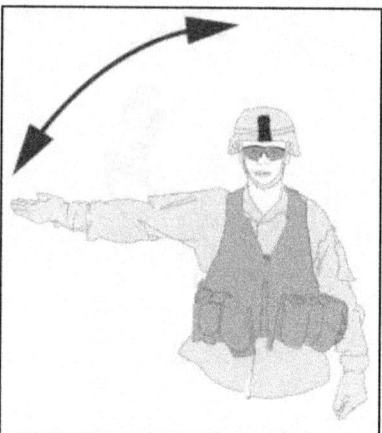

Figure 1-14. Contact right

1-18. To signal "air attack," bend the arms at a 45-degree angle. The forearms are crossed. Repeat. (See figure 1-15.)

Figure 1-15. Air attack

1-19. To signal "chemical, biological, radiological and nuclear attack," extend the arms and fists. Bend the arms to the shoulders. Repeat. (See figure 1-16.)

Figure 1-16. Chemical, biological, radiological, and nuclear attack

1-20. To signal "fix bayonets," simulate the movement of the right hand in removing the bayonet from the scabbard and fixing it on the rifle. (See figure 1-17.)

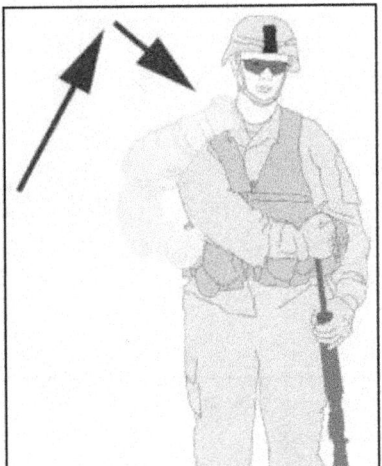

Figure 1-17. Fix bayonets

1-21. To signal "enemy in sight," hold the rifle in the ready position at shoulder level. Point the rifle in the direction of the enemy. Alternately, use the non-firing hand, point index finger at the enemy and thumb pointing down. All other fingers will be curled. (See figure 1-18.)

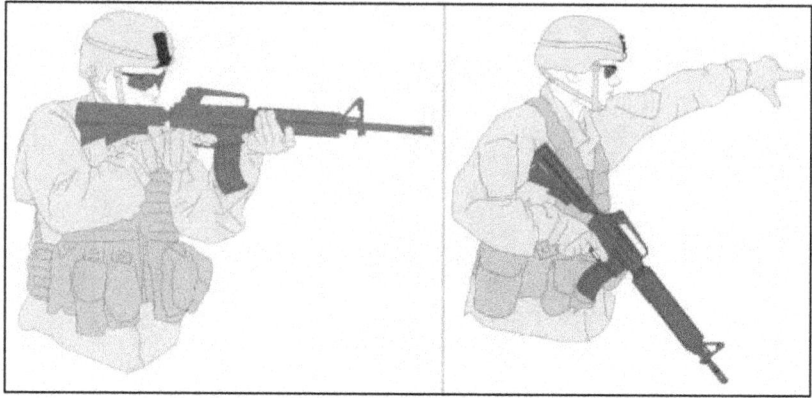

Figure 1-18. Enemy in sight

1-22. To signal "quick time," extend the arm horizontally sideward, palm to the front, and wave the arm slightly downward several times, keeping the arm straight. Do not move the arm above the horizontal. (See figure 1-19.)

Figure 1-19. Quick time

Hand and Arm Signals for Dismounted Operations

1-23. To signal "take cover," extend the arm at a 45-degree angle from the side, above the horizontal, palm down, and then lower the arm to the side. (See figure 1-20.)

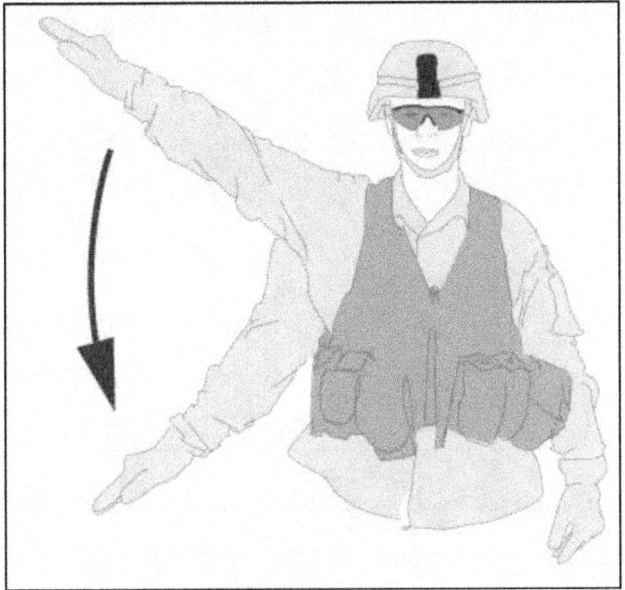

Figure 1-20. Take cover

1-24. To request a "map check," point at the palm of one hand with the index finger of the other hand. (See figure 1-21.)

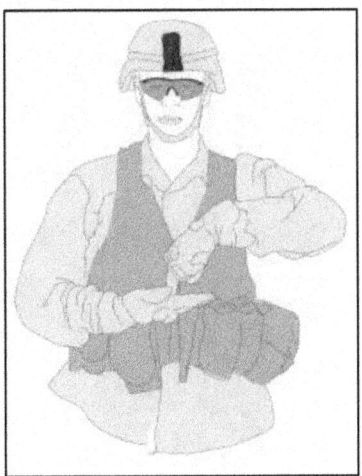

Figure 1-21. Map check

1-25. To signal "halt," raise hand to head level, fingers extended and joined. (See figure 1-22.)

Figure 1-22. Halt

1-26. To signal "take a knee," following the signal for halt, lower hand from the halt position to waist level, palm facing down, fingers extended and joined. (See figure 1-23.)

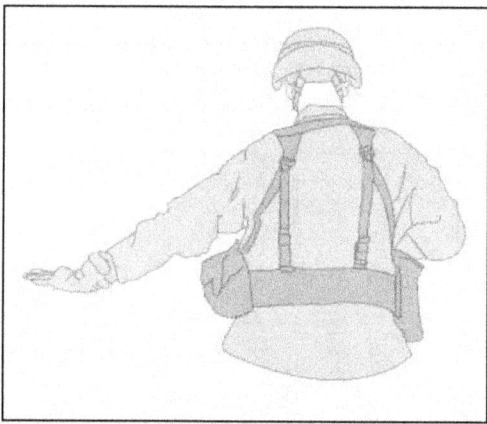

Figure 1-23. Take a knee

Hand and Arm Signals for Dismounted Operations

1-27. To signal "move to the prone," following the signal for halt or take a knee, lower the arm past waist level, palm facing down, then move forearm from waist level to a position lower than waist level repeatedly. Palm will be toward ground; fingers will be extended and joined. (See figure 1-24.)

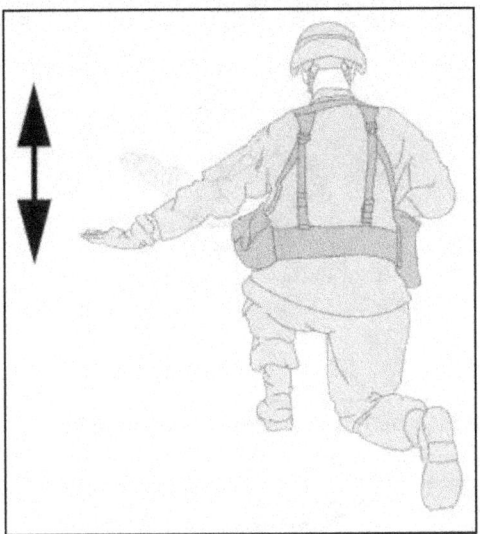

Figure 1-24. Move to the prone

1-28. To request a pace count, tap the heel of boot repeatedly with an open hand. (See figure 1-25.)

Figure 1-25. Pace count

1-29. To call the radiotelephone operator forward, raise the hand to the ear with the thumb and little finger extended. (See figure 1-26.)

Figure 1-26. Radiotelephone operator forward

1-30. To request a headcount, tap the back of the helmet repeatedly with an open hand. (See figure 1-27.)

Figure 1-27. Head count

1-31. To signal "danger area," draw the right hand, palm down, across the neck in a throat-cutting motion from left to right. (See figure 1-28.)

> **Note.** This movement is the same as figure 2-18, Stop Engines, on page 2-12. The difference in meanings is understood from the context in which it is used.

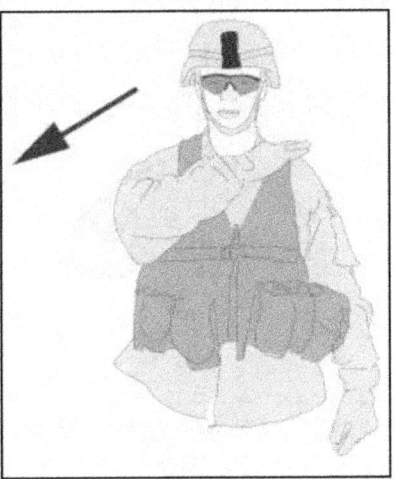

Figure 1-28. Danger area

1-32. To signal "freeze," raise the fist to head level. (See figure 1-29.)

Figure 1-29. Freeze

Chapter 1

1-33. To signal "move the platoon leader to the front," use non-firing hand and place index finger to the rim of headgear. (See figure 1-30.)

Figure 1-30. Move platoon leader to the front

1-34. To signal "move the platoon sergeant to the front," use non-firing hand to grab the collar. In the absence of a uniform collar, simulate the movement. (See figure 1-31.)

Figure 1-31. Move the platoon sergeant to the front

1-35. To signal "move the squad leader forward," use non-firing hand and place finger on upper arm. Place the number of fingers that correspond to the appropriate squad leader. For example, two fingers means, "second squad leader forward." (See figure 1-32.)

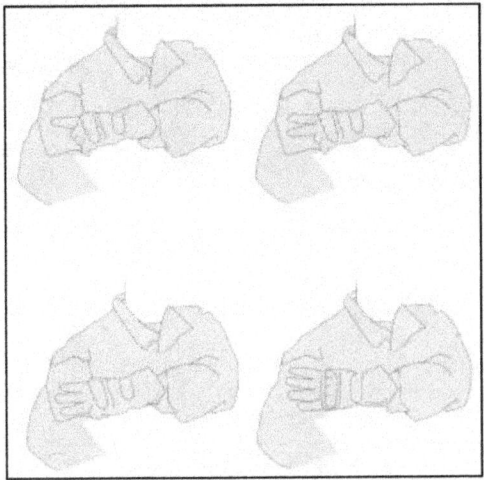

Figure 1-32. Move squad leader forward

1-36. All number signals will be given with the non-firing hand. For numbers greater than nine, the signal will be given as individual digits in the number. For example, the number 27 will be represented by the signal for the number, 2, followed the signal for the number, 7. (See figure 1-33.)

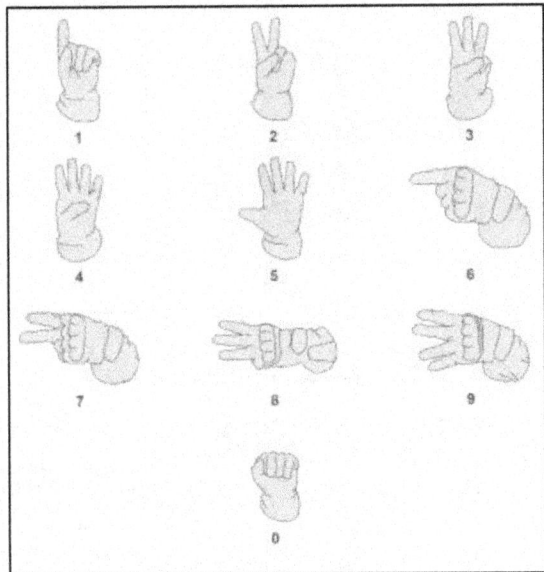

Figure 1-33. Number signals

1-37. To signal "stop, look, listen, smell (SLLS)," raise the open palm of the non-firing hand to ear level. (See figure 1-34.)

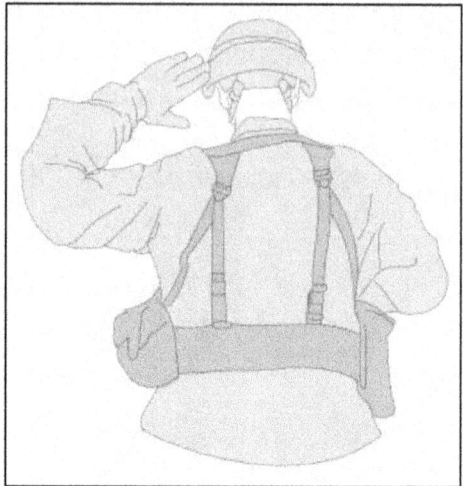

Figure 1-34. Stop, look, listen, smell

1-38. To signal "message acknowledged," hold the fist out and thumb up. (See figure 1-35.)

Figure 1-35. Message acknowledged

SIGNALS FOR CREW-SERVED WEAPONS

1-39. Members of crew-served weapons must communicate. Often, this is in environments where visual signals are the best means of transmitting information. (See figures 1-36 through 1-40 on page 1-18 through page 1-21.)

1-40. To signal "commence firing," extend the arm in front of the body, palm down, and move it through a wide horizontal arc several times. For machine guns, when giving the signal again, moving the arm faster means to change to the next higher rate of fire. To slow the rate of fire, move the arm slower. This signal is primarily used for direct fire weapons. (See figure 1-36.)

Figure 1-36. Commence firing or change rate of fire

1-41. To signal the gunner to "change direction" or "change elevation," the signaler will move his hand and arm in new direction and indicates the amount of change by the number of fingers shown to the gunner. Each finger shown represents a 1 mil or 1 meter change in direction. The signaler will extended his hand repeatedly to indicate the total amount of change. For example: the signaler represents UP EIGHT meters by moving his arm upward with five fingers showing then moving his arm upward again with three fingers showing. (See figure 1-37.)

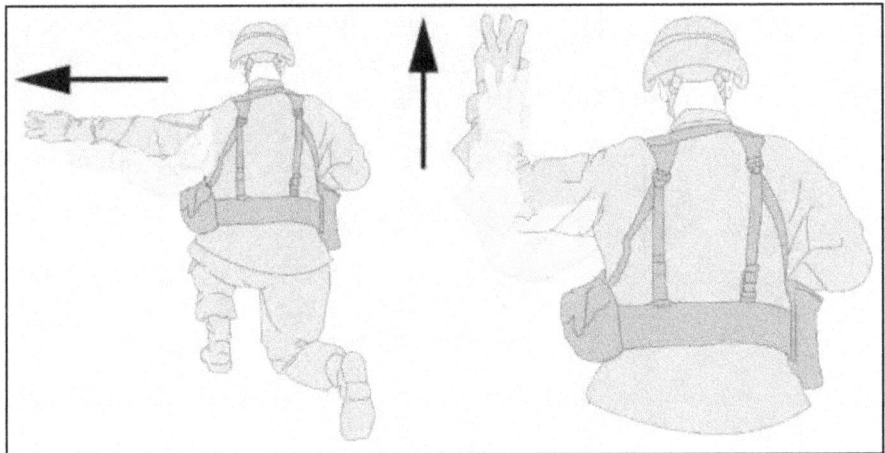

Figure 1-37. Change direction or elevation.

1-42. To signal "move over" or "shift fire." raise the hand (on the side toward the new direction) and move it across the body to the opposite shoulder, palm to the front; then swing the arm in a horizontal arc, extending the arm and hand to point in the new direction. For slight changes in direction, move the hand from the final position to the desired direction of movement. (See figure 1-38.)

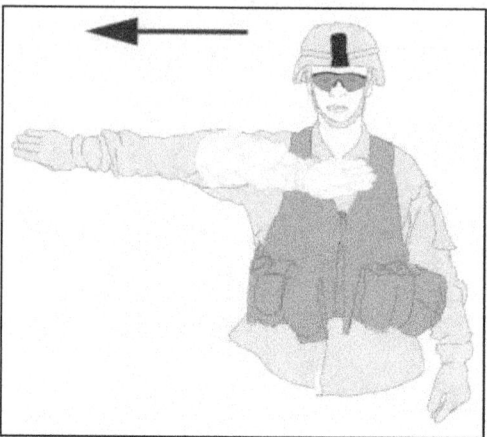

Figure 1-38. Move over or shift fire

1-43. To signal "cease firing," raise the hand in front of the forehead, palm to the front, and swing the hand and forearm up and down several times in front of the face. (See figure 1-39.)

Figure 1-39. Cease firing

1-44. To signal "out of action," strike the fist of one hand several times in rapid succession against the palm of the other hand. (See figure 1-40.)

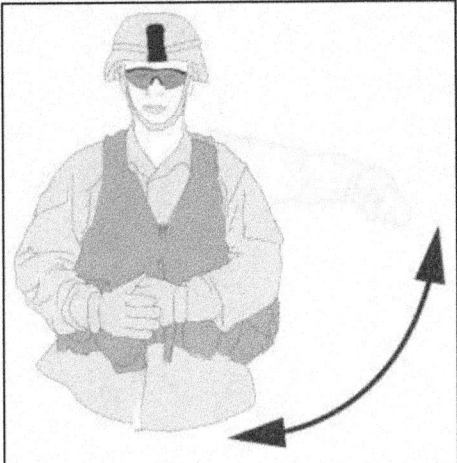

Figure 1-40. Out of action

This page intentionally left blank.

Chapter 2
Hand-and-Arm Signals for Ground Vehicles

GENERAL

2-1. Signals illustrated from the hatch or top of a tactical vehicle are for illustration only or to be given while the vehicle is stationary. As per Army Regulation 385-10, personnel will not expose more than their head and shoulders (name tag defilade) except when actively engaging targets with the vehicle mounted weapons systems.

SIGNALS FOR MECHANIZED MOVEMENT TECHNIQUES

2-2. Signals for movement techniques are used by mechanized units to indicate which manner of traversing terrain will be used by a unit. (See figures 2-1 through 2-21 on page 2-1 through page 2-11.)

2-3. To signal "wedge formation," extend the arms downward and to the sides at a 45-degree angle below the horizontal, palms to the front. (See figure 2-1.)

Figure 2-1. Wedge formation

2-4. To signal "vee formation," raise the arms and extend them 45 degrees above the horizontal. (See figure 2-2.)

Figure 2-2. Vee formation

2-5. To signal "line formation," extend the arms parallel to the ground. (See figure 2-3.)

Figure 2-3. Line

2-6. To signal "coil formation," raise one arm above the head and rotate it in a small circle. (See figure 2-4.)

Figure 2-4. Coil

2-7. To signal "echelon left," extend the right arm and raise it 45 degrees above the shoulder. Extend the left arm 45 degrees below the horizontal and point toward the ground. (See figure 2-5.)

Figure 2-5. Echelon left

Chapter 2

2-8. To signal "echelon right," extend the left arm and raise it 45 degrees above the shoulder. Extend the right arm 45 degrees below the horizontal and point toward the ground. (See figure 2-6.)

Figure 2-6. Echelon right

2-9. To signal "staggered column," extend the arms so that upper arms are parallel to the ground and the forearms are perpendicular. Raise the arms so they fully extended above the head. Repeat. (See figure 2-7.)

Figure 2-7. Staggered column formation

Hand and Arm Signals for Wheeled and Mounted Operations

2-10. To signal "column formation," raise and extend the arm overhead. Move it to the right and left. Continue until the formation is executed. Alternately, to signal form a file or column, move the non-firing hand to touch the rim of the headgear directly in front of the face. Fingers will be extended and joined. (See figure 2-8.)

Figure 2-8. Column formation

2-11. To signal form a "herringbone formation," extend the arms parallel to the ground. Bend the arms until the forearms are perpendicular. Repeat. (See figure 2-9.)

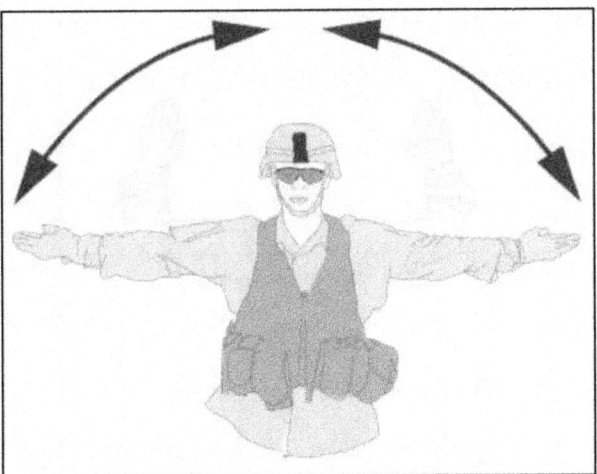

Figure 2-9. Herringbone formation

Chapter 2

2-12. To signal "traveling," extend the arm overhead and swing it in a circle from the shoulder. (See figure 2-10.)

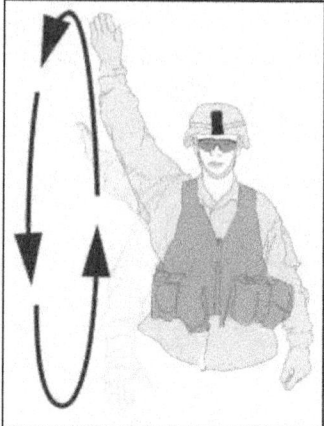

Figure 2-10. Traveling

2-13. To signal "traveling overwatch," extend both arms and raise them up and down. (See figure 2-11.)

Figure 2-11. Traveling overwatch

Hand and Arm Signals for Wheeled and Mounted Operations

2-14. To signal "bounding overwatch," extend one arm to a 45-degree angle. Bend the arm and tap the helmet. Repeat. (See figure 2-12.)

Figure 2-12. Bounding overwatch

2-15. To signal "fire," drop the arm sharply from the vertical position (usually from the Are you ready signal position) to the side. When a single weapon (of a group) is to be fired, point, with the arm extended, to that particular weapon, and then drop the arm sharply to the side. Usually the signal is used as a fire command for indirect fire weapons. (See figure 2-13.)

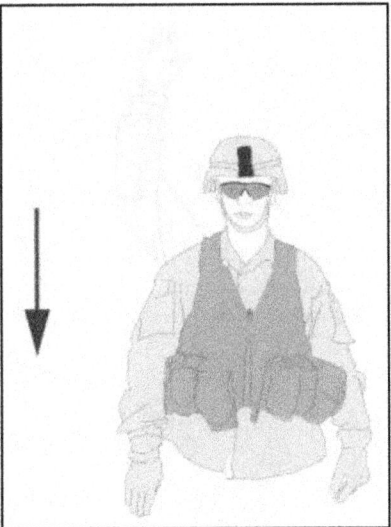

Figure 2-13. Fire

2-16. To signal "move to the left," extend the arm to the left and raise it up and down. (See figure 2-14.)

Figure 2-14. Move to the left

2-17. To signal "move to the right," extend the arm to the right and raise it up and down. (See figure 2-15.)

Figure 2-15. Move to the right

Hand and Arm Signals for Wheeled and Mounted Operations

2-18. To signal "advance," "move out," or "follow me," face the direction of movement; hold the arm extended to the rear; swing the arm overhead and forward in the direction of movement (hold at the horizontal), palm down. (See figure 2-16.)

Figure 2-16. Advance, move out or "follow me"

2-19. To signal "dismount," extend the arms, make two or three movements up and down, hands open toward ground. (See figure 2-17.)

Figure 2-17. Dismount

2-20. To signal "stop," clasp the hands together, palms facing, at chin level. (See figure 2-18.)

Note. For alternate signal to stop vehicles, see figure 2-28 on page 2-15.

Figure 2-18. Stop

2-21. To signal "button up," place both hands, one on top of the other, palms down, on top of the helmet. The arms are back and in the same plane as the body. For "unbutton," give the button up signal, then separate the hands, moving them to each side in a slicing motion; repeat. (See figure 2-19.)

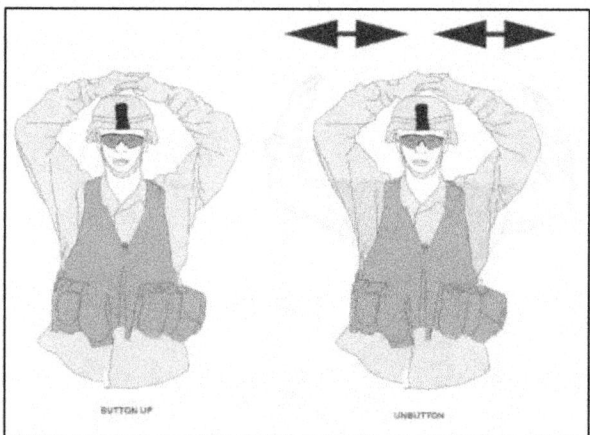

Figure 2-19. Button up or unbutton

Hand and Arm Signals for Wheeled and Mounted Operations

2-22. To signal "open up," extend the arms overhead, palms inward, then slowly lower arms to a horizontal position. (See figure 2-20.)

Figure 2-20. Open up

2-23. To signal "close up," extend both arms parallel to the ground, palms uppermost, then move the arms upward and inward toward the head. (See figure 2-21.)

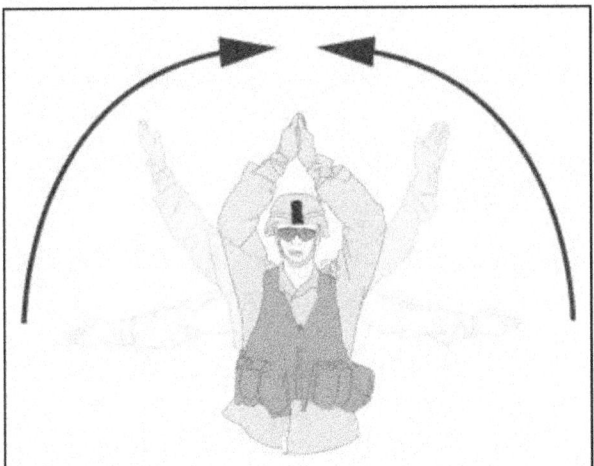

Figure 2-21. Close up

SIGNALS TO CONTROL VEHICLE DRIVERS AND/OR CREWS

2-24. These are the hand and arm and light signals used to guide and direct vehicles. Flashlights are used at night to direct vehicles. Blue filters should be used whenever possible to preserve the driver's night vision. Chemical lights also can be used and have less effect on the driver's night vision. (See figures 2-22 through 2-36 on page 2-11 through page 2-19.)

2-25. To signal "attention," extend the arm sideways, slightly above horizontal; palm to the front; wave the arm to and from the head several times. (See figure 2-22.)

Figure 2-22. Attention

2-26. To signal "I am ready," or ask if a driver or vehicle is "ready," extend the arm toward the person being signaled; then raise the arm slightly above horizontal, palm outward. (See figure 2-23.)

Figure 2-23. I am ready or ready to move or are you ready?

2-27. To signal "mount," make two or three movements upward with the open hand, palm facing upward. (See figure 2-24.)

Figure 2-24. Mount

2-28. To signal "disregard previous command" or "remain in place," raise both arms and cross wrists above the head, palms facing out. (See figure 2-25.)

Figure 2-25. Disregard previous command or remain in place

2-29. To signal "I do not understand," raise both arms sideward to the horizontal; bend both arms at the elbows and place both hands across the face, palms to the front. (See figure 2-26.)

Figure 2-26. I do not understand

2-30. To signal "start engine" or "prepare to move," during daylight hours simulate cranking of engines by moving the arm, with the fist, in a circular motion at waist level; at night, move a light to describe a horizontal figure 8 in a vertical plane in front of body. (See figure 2-27.)

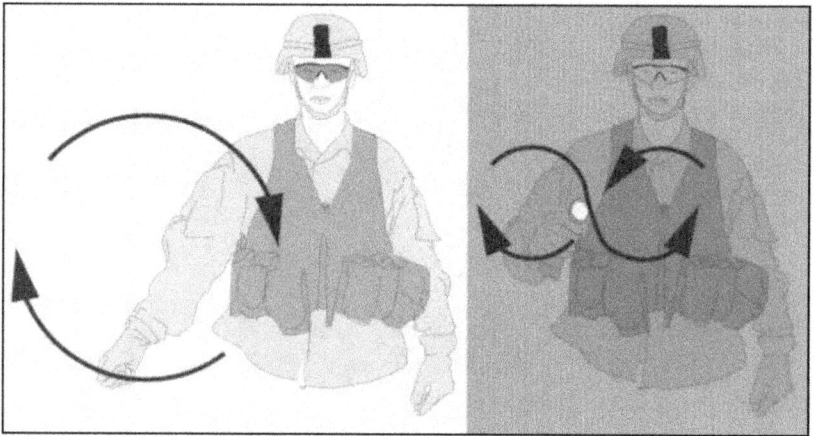

Figure 2-27. Start engine or prepare to move

Hand and Arm Signals for Wheeled and Mounted Operations

2-31. To signal "halt" or "stop," during daylight hours, raise the hand upward to the full extent of the arm, palm to the front and hold that position until the signal is understood. At night, move a light horizontally back and forth several times across the path of approaching traffic to stop vehicles. Use the same signal to stop engines. (See figure 2-28.)

Note. For an alternate signal to stop vehicles, see figure 2-18 on page 2-10.

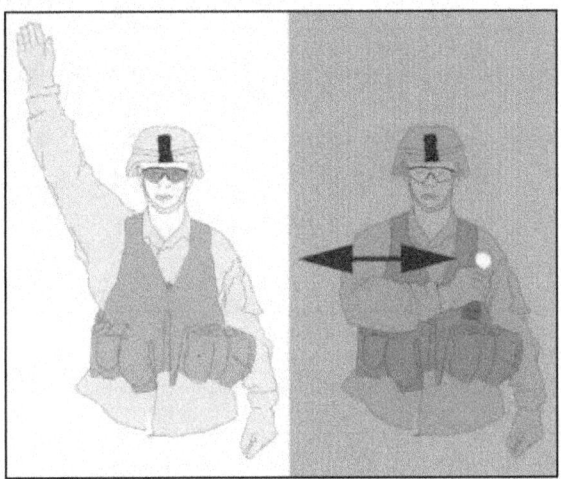

Figure 2-28. Halt or stop

2-32. To signal "increase speed," during daylight hours, raise the fist to shoulder level; thrust the fist upward to the full extent of the arm and back to shoulder level rapidly several times. At night move a light vertically several times in front of the body. (See figure 2-29.)

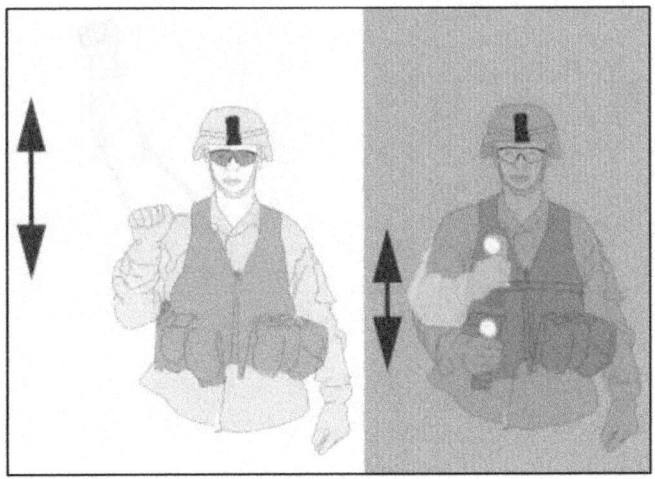

Figure 2-29. Increase speed

Chapter 2

2-33. To signal "right turn" or "left turn," during daylight hours, extend the arm horizontally to side, palm outward. At night, rotate a light to describe a circle 12 to 18 inches in diameter in the direction of the turn. (See figure 2-30.)

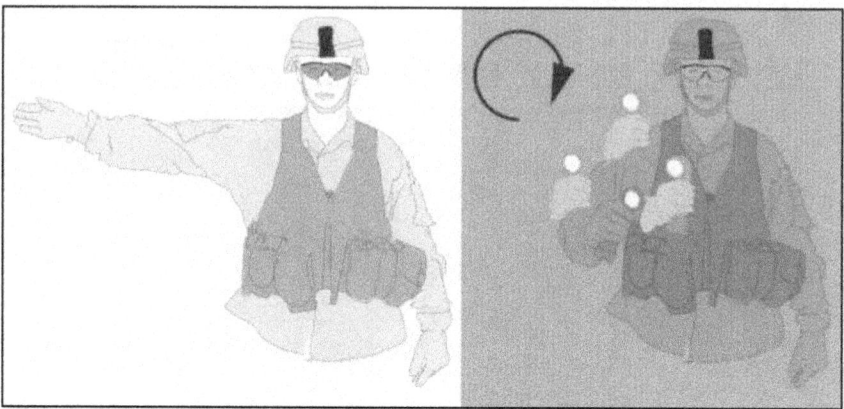

Figure 2-30. Right or left turn

2-34. To signal "slow down," during daylight hours, extend the arm horizontally sideward, palm facing downward; wave the arm slightly downward several times, keeping the arm straight. Do move the arm above horizontal. (See figure 2-31.)

Figure 2-31. Slow down

2-35. To signal "move forward," move the hands and forearms backward and forward, palms toward the chest. (See figure 2-32.)

Figure 2-32. Move forward

2-36. To signal "move in reverse," during daylight hours, face the vehicle(s) (unit) being signaled, raise the hands to shoulder level, palms to the front. Move the hands forward and backward. At night, hold a light a shoulder level; blink it several times toward the vehicle(s). (See figure 2-33.)

Figure 2-33. Move in reverse (for stationary vehicles)

2-37. To signal "close distance between vehicles and stop," face the vehicle(s) being signaled, extend the forearms to the front, palms inward and separated (width of the shoulder(s). Bring the palms together as the vehicle(s) approaches. The vehicle(s) must stop when the palms come together. (See figure 2-34.)

Figure 2-34. Close distance between vehicles and stop

2-38. To signal "stop engines," extend the arm parallel to ground, hand open, and move the arm across the body, in a throat-cutting action. (See figure 2-35.)

Figure 2-35. Stop engines

2-39. To signal "neutral steer" for track vehicles, cross the wrists at the throat; point the index finger in direction of steer. Make a fist of the other hand. (See figure 2-36.)

Figure 2-36. Neutral steer (track vehicles)

TRAFFIC CONTROL

2-40. These signals normally are used by authorized officials (civilian and military police, and personnel at traffic control points) to direct traffic. At night, these signals are given with a flashlight or a lighted wand. (See figures 2-37 through 2-41 on page 2-19 through page 2-21.)

2-41. To signal "left and right traffic stop," stand facing traffic with the arms raised, palms open, in the same plane as the shoulders. (See figure 2-37.)

Figure 2-37. Left and right traffic stop

2-42. To signal "front traffic stop," stand facing traffic with arm raised, palm open. (See figure 2-38.)

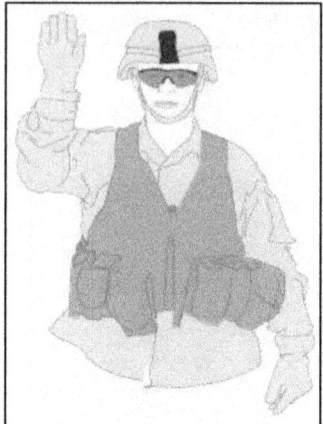

Figure 2-38. Front traffic stop

2-43. To signal "rear traffic stop," stand with the back to traffic, the arm raised, palm open. Rotate the upper body so the palm faces traffic. (See figure 2-39.)

Figure 2-39. Rear traffic stop

2-44. To signal "traffic from the right, GO," stand with the right side facing traffic, left arm extended, palm open. The right arm is parallel to the ground and bent with the palm at shoulder level. (See figure 2-40.)

Figure 2-40. Traffic from right, GO

2-45. To signal "traffic from the left, GO," stand with the left side facing traffic, right arm extended, palm open. The left arm is parallel to the ground with the palm at shoulder level. (See figure 2-41.)

Figure 2-41. Traffic from left, GO

CONVOY CONTROL

2-46. In addition to traffic control personnel, convoy commanders can use hand and arm signals to convey messages (see figures 2-42 through 2-45 on page 2-22 and page 2-23).

Chapter 2

2-47. Signals given by the convoy commander will be given out of the passenger side window or relayed to personnel operating a mounted weapons system. The driver should be the last resort for relaying hand and arm signals to the convoy due to safety considerations.

2-48. To order drivers to "open up" or "increase the distance between vehicles," extend the left arm horizontally to the side, palm to the front, then move the arm downward to a 45-degree angle below horizontal. Repeat several times. (See figure 2-42.)

Figure 2-42. Open up or increase the distance between vehicles

2-49. To order drivers to "close up" or "decrease the distance between vehicles," extend the left arm sideward to the horizontal, palm up, and raise it to the vertical. Repeat several times. (See figure 2-43.)

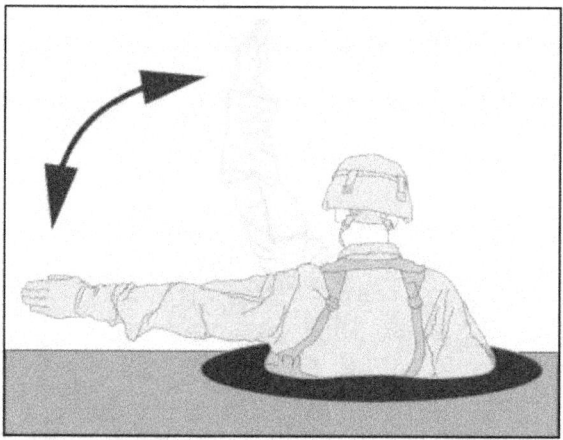

Figure 2-43. Close up or decrease the distance between vehicles

Hand and Arm Signals for Wheeled and Mounted Operations

2-50. To order another vehicle or vehicles to "pass and keep going," extend the left arm horizontally to the side, palm to the front, and describe large circles to the front by rotating the arm clockwise from the elbow. (See figure 2-44.)

Figure 2-44. Pass and keep going

2-51. To order a vehicle to "move in reverse," face the unit being signaled and raise the hand to shoulder level in front of the body, palm to the front; extend the arm forward to its full extent in a pushing motion, keeping the palm to the front. (See figure 2-45.)

Note. This is done when the commander's vehicle has halted.

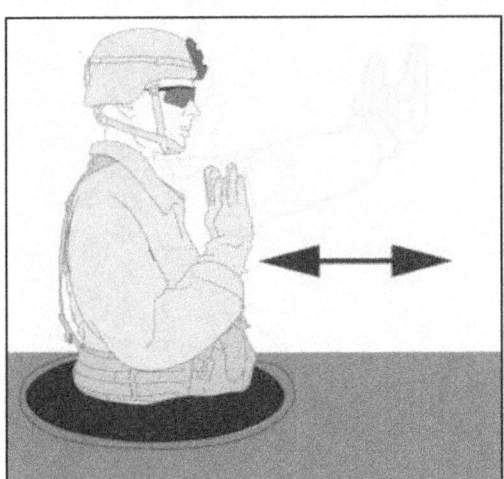

Figure 2-45. Move in reverse

FLAG SIGNALS

2-52. Flags are issued to armored and mechanized units for control purposes and as an alternate means of communication within these units. Each combat vehicle is equipped with a flag set consisting of one red, one yellow, and one green flag. Flag signals may be given by using a single flag or a combination of two or three flags, according to a prearranged code. Flag signals, when understood, are repeated and executed at once (see figures 2-46 through 2-51 on page 2-24 through page 2-27).

2-53. Flags are used to—
- Mark vehicle positions. For example, a quartering party member uses colored flags in an assembly area to mark positions.
- Identify disabled vehicles.
- Warn friendly elements of an advancing enemy. For example, an observation post uses a flag to signal a platoon to move to its fighting position.
- Control movement. Flags serve as an extension of hand and arm signals when distances between vehicles become too great.

2-54. When used alone, flag colors have the following meanings:
- Red – DANGER, or ENEMY IN SIGHT.
- Green – ALL CLEAR, READY, or UNDERSTOOD.
- Yellow – DISREGARD, or VEHICLE OUT OF ACTION.

2-55. During periods of limited visibility flashlights with colored filters or colored chemical lights may be substituted for flags.

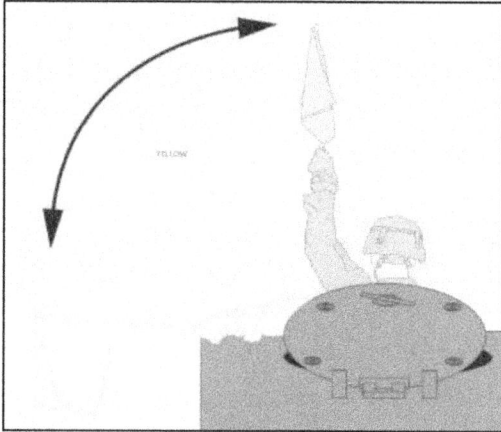

Figure 2-46. Mount

Hand and Arm Signals for Wheeled and Mounted Operations

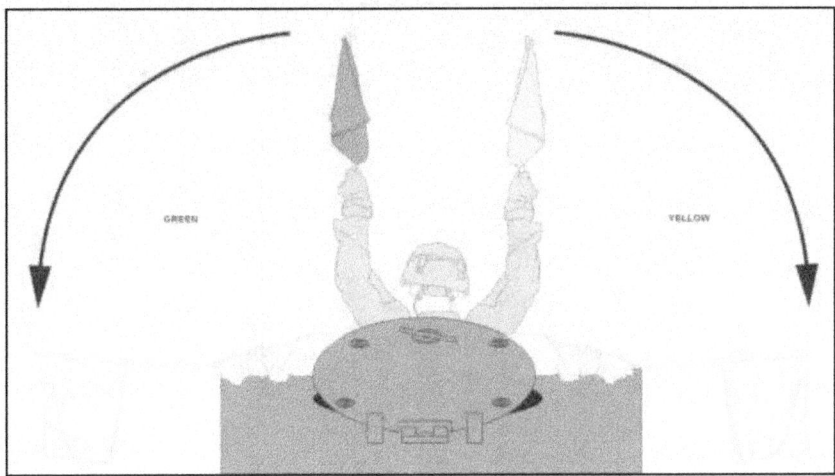

Figure 2-47. Dismount

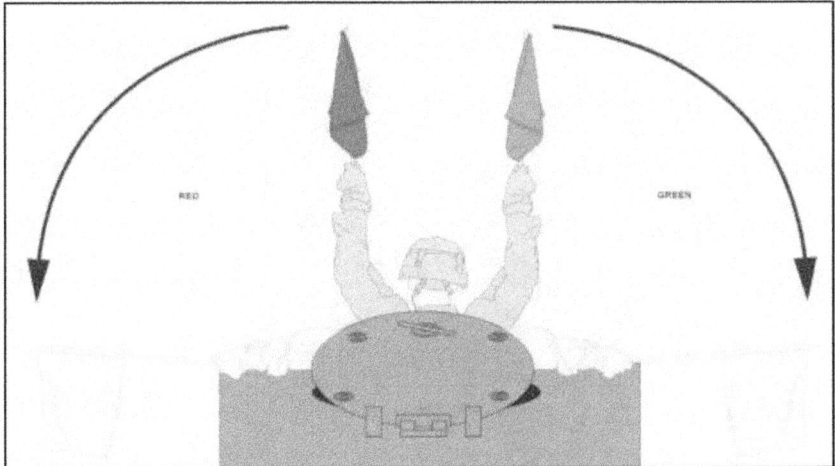

Figure 2-48. Dismount and assault

Chapter 2

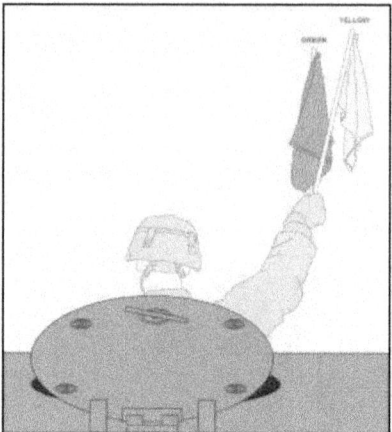

Figure 2-49. Assemble or close

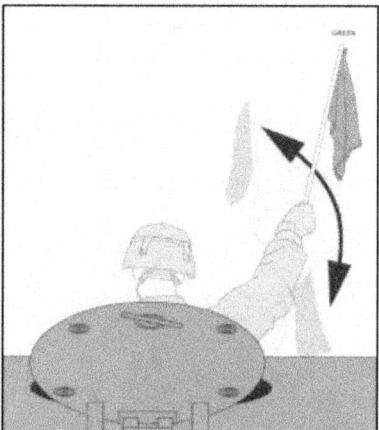

Figure 2-50. Move out

Hand and Arm Signals for Wheeled and Mounted Operations

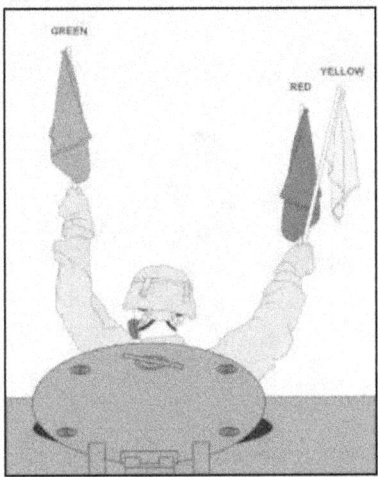

Figure 2-51. Chemical, biological, radiological, and nuclear hazard present

FIRING RANGE FLAG SIGNALS

2-56. Signal flags are used on firing ranges for tanks or fighting vehicles to indicate the status of the range and the status of the individual vehicle. A red flag at the control point indicates that firing may be conducted, while a green flag indicates that it may not. (See figures 2-52 through 2-56 on page 2-27 through 2-29.)

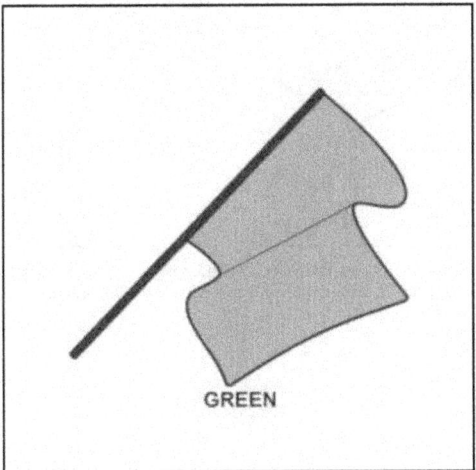

Figure 2-52. All weapons clear (guns elevated)

Chapter 2

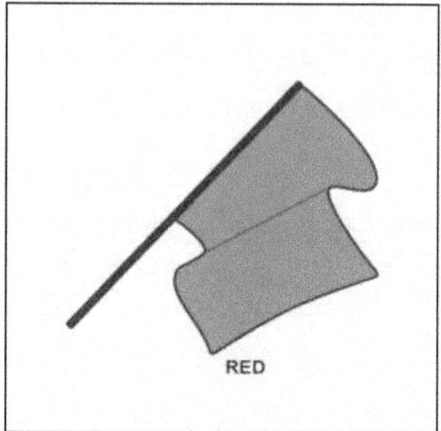

Figure 2-53. Conducting live fire or "hot gun"

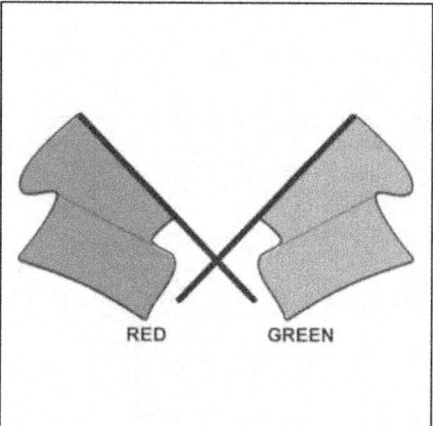

Figure 2-54. Conducting prepare-to-fire or non-firing exercises
(Ammunition is uploaded and the system is on safe.)

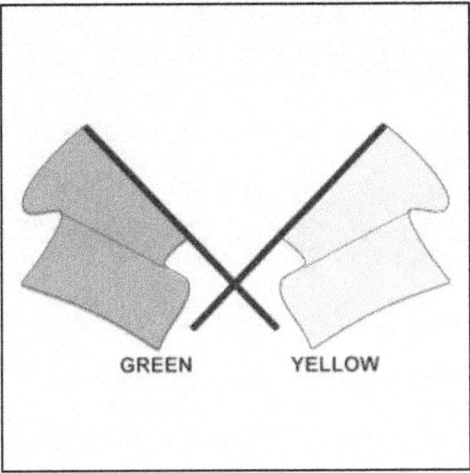

Figure 2-55. Malfunction—weapons clear

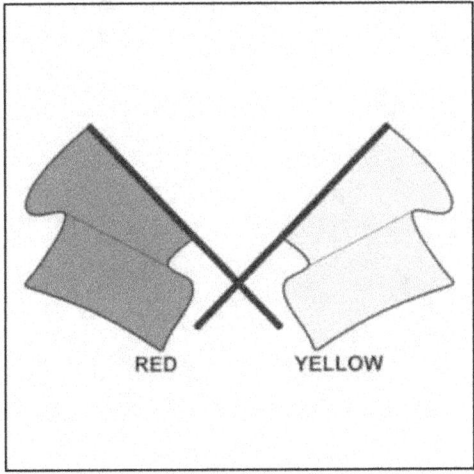

Figure 2-56. Malfunction—weapons loaded

This page intentionally left blank.

Chapter 3

Hand and Arm Signals for Aircraft

GENERAL

3-1. Personnel on the ground and aviation assets need to communicate. There will be times when radios and digital systems cannot be used and visual signals must be used. Therefore, systems of standard visual signals have been developed to allow ground-to-air communication. These systems include hand and arm signals used by ground forces to direct helicopters in direct support; devices that can be used to communicate with aircraft; and ground-to-air emergency signals and codes.

COMMON HAND AND ARM GROUND SIGNALS

3-2. Helicopters and fixed-wing aircraft are often used to support ground forces by moving supplies and personnel. Often, pathfinder personnel will not be available to direct aircraft in support of these efforts. Therefore, the responsibility to guide aircraft will fall upon the ground forces. To be prepared for this effort, the Soldier must know these general signals (see figures 3-1 through 3-22 on page 3-1 through page 3-12).

3-3. When directing a taxiing helicopter, the signalman's position is slightly to the right, in full view of the pilot, and at a safe distance of no less than 40 meters (or, no closer 20 meters during slingload operations). These positions are used during day and night operations. The signalman never stands in front of an armed helicopter or fixed-wing aircraft. (See figure 3-1 and figure 3-2.)

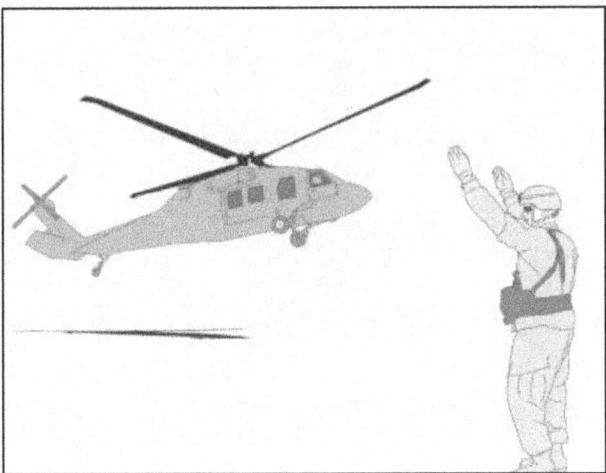

Figure 3-1. Helicopters (rotary wing)

Chapter 3

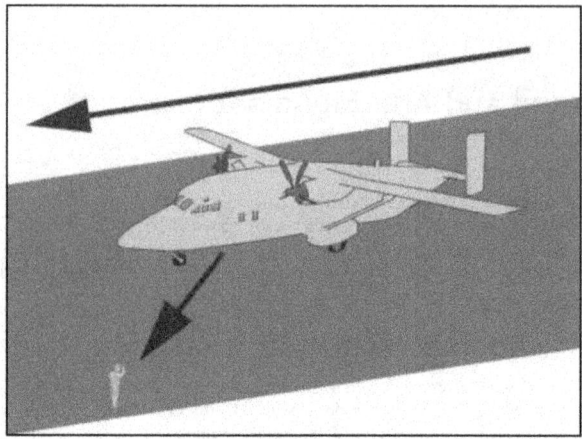

Figure 3-2. Fixed-wing aircraft

3-4. To signal "assume guidance," extend arms above the head in a vertical position, palms facing forward. (See figure 3-3.)

Figure 3-3. Assume guidance

Hand and Arm Signals for Aircraft

3-5. To signal "cut engine" or "stop rotor," either arm, with the shoulder, palm down. Draw the extended hand across the neck in a throat-cutting motion. If a specific engine (rotor) or a multi-engine (rotor) aircraft is to be shut down, execute the signal and point with the other hand to the appropriate engine (rotor). (See figure 3-4.)

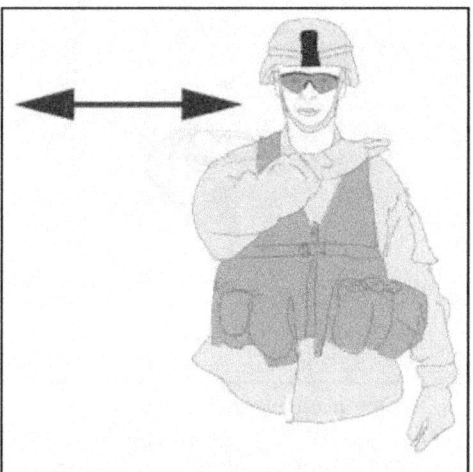

Figure 3-4. Cut engine(s) or stop rotor(s)

3-6. To signal "emergency," to the aircraft crew, the hand and arm will move in a horizontal "infinity" symbol with the palm facing the aircraft. (See figure 3-5.)

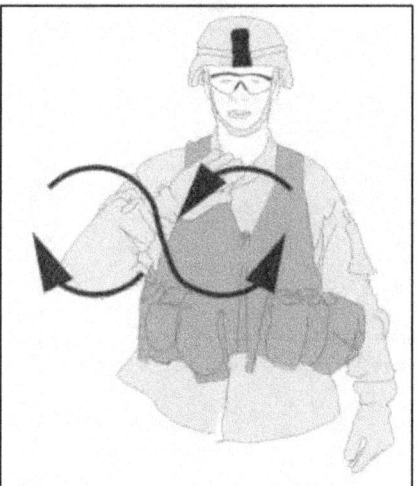

Figure 3-5. Emergency signal

Chapter 3

3-7. To signal "proceed right to next signalman," hold the left arm down. Extend the right arm across the body to indicate the direction to the next signalman. (See figure 3-6.)

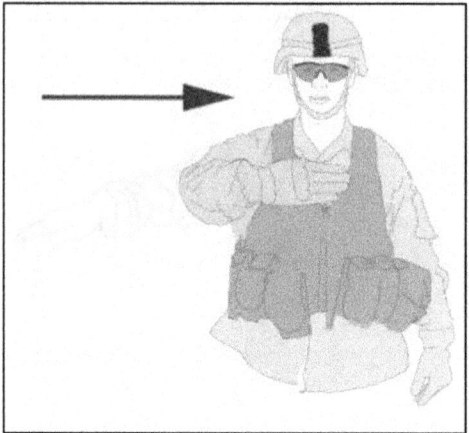

Figure 3-6. Proceed right to next signalman

3-8. To signal "proceed left to the next signalman," hold the right arm down. Extend the left arm across the body to indicate the direction to the next signalman. (See figure 3-7.)

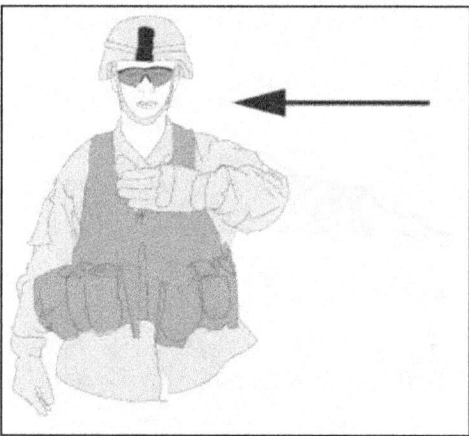

Figure 3-7. Proceed left to next signalman

3-9. To signal depart, make an overhead circular motion with the right hand, ending it in a throwing motion in the direction of lift-off (takeoff). (See figure 3-8.)

Figure 3-8. Depart

3-10. To signal "go around" or "do not land," cross the arms repeatedly overhead. (See figure 3-9.)

Figure 3-9. Go around, do not land

3-11. To signal "land," extend the crossed arms downward in front of the body. (See figure 3-10.)

Figure 3-10. Land

3-12. To signal "stop," cross the arms above the head, palms forward. (See figure 3-11.)

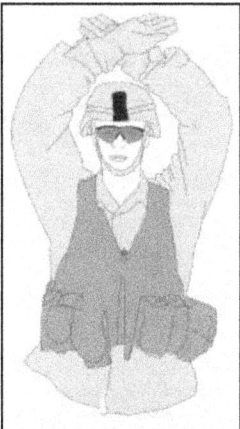

Figure 3-11. Stop

Hand and Arm Signals for Aircraft

3-13. To signal "spot turn," move the hand upward and backward, from a horizontal position, to indicate direction of tail movement. Point the other hand toward the center of the spot turn. The signalman must remain in full view of the pilot. (See figure 3-12.)

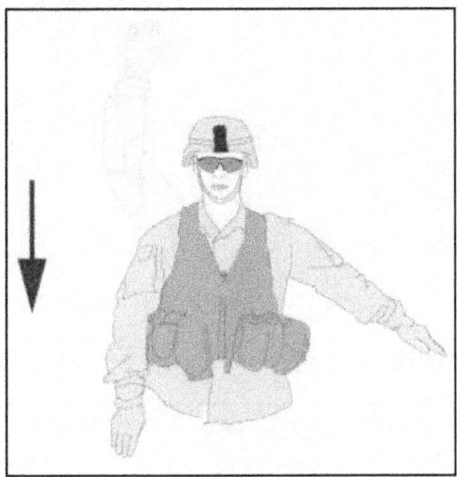

Figure 3-12. Spot turn

3-14. To signal "move right," extend the left arm horizontally to the side in the direction of movement; swing the right arm over the head in the same direction (repeat movement). (See figure 3-13.)

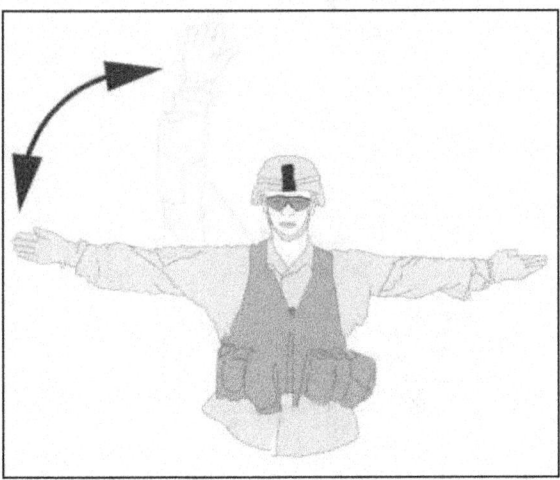

Figure 3-13. Move right

Chapter 3

3-15. To signal "move left," extend the right arm horizontally to the side in the direction of movement; swing the left arm over the head in the same direction (repeat movement). (See figure 3-14.)

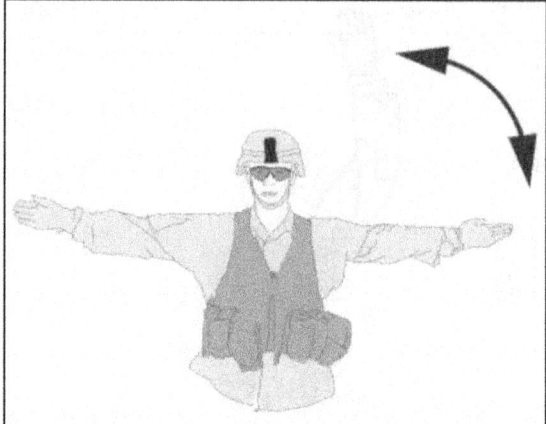

Figure 3-14. Move left

3-16. To signal "move ahead," extend the arms slightly away from the side, palms to the rear, and repeatedly move them upward and backward (from shoulder height). This signal is used to indicate short distances. (See figure 3-15.)

Figure 3-15. Move ahead

Hand and Arm Signals for Aircraft

3-17. To signal "move rearward," place arms by the sides, palms to the front. Sweep the arms forward and upward repeatedly, level with the shoulders. (See figure 3-16.)

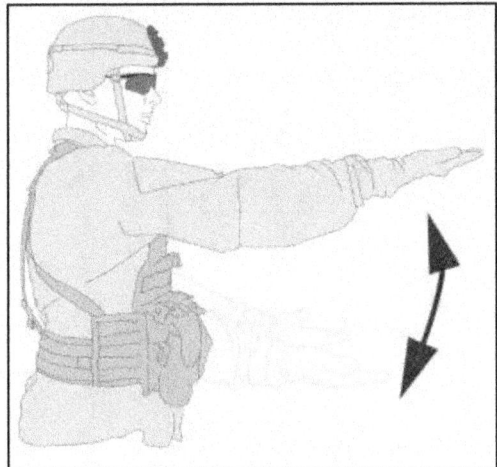

Figure 3-16. Move rearward

HAND AND ARM SIGNALS FOR ROTARY-WING AIRCRAFT

3-18. The following signals are used when guiding rotary wing aircraft and during sling load operations.

3-19. To signal "load has not been released," bend the left arm and fist horizontally across the chest (knuckles down); point the open, right hand up to the center of the left fist. (See figure 3-17.)

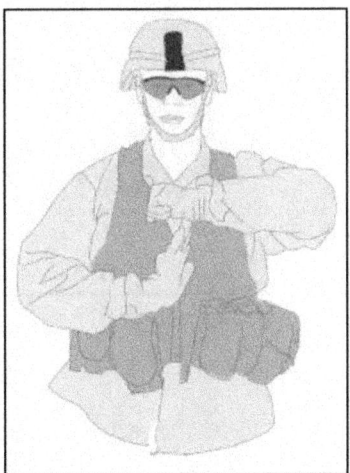

Figure 3-17. Load has not been released

Chapter 3

3-20. To signal "hook up complete," move the fist up and down, making contact with the other fist, which is stationary and on top of the helmet. (See figure 3-18.)

Figure 3-18. Hookup complete

3-21. To signal "release," extend the left arm horizontally with the fist toward the load while the right arm makes a horizontal, slicing motion under the left arm, palm down. (See figure 3-19.)

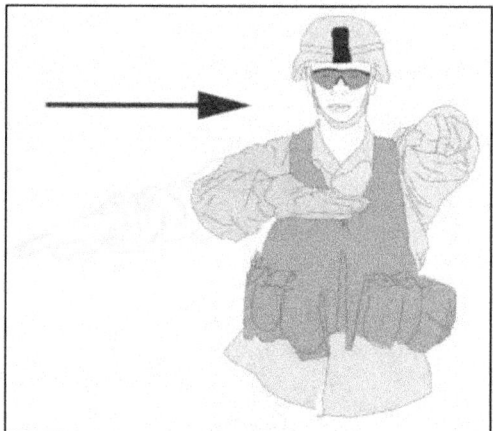

Figure 3-19. Release

Hand and Arm Signals for Aircraft

3-22. To signal "move downward," extend the arms horizontally to the sides, beckon downward, palms down. (See figure 3-20.)

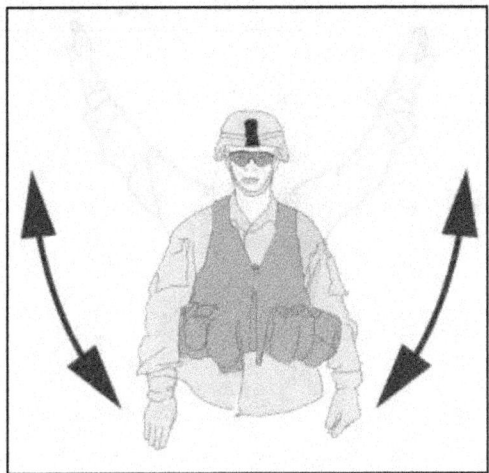

Figure 3-20. Move downward

3-23. To signal "move upward," extend the arms horizontally to the sides, beckon upward, palms up. (See figure 3-21.)

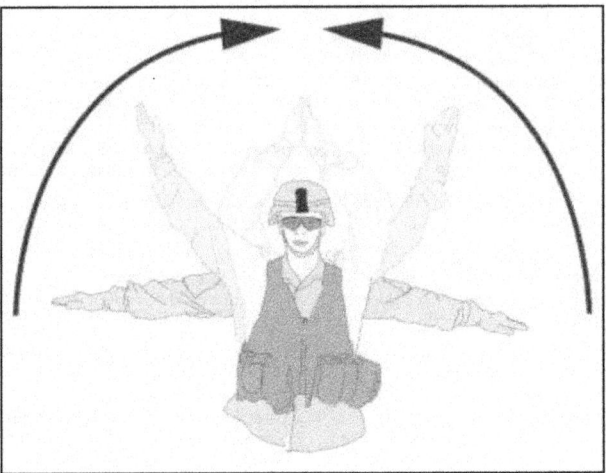

Figure 3-21. Move upward

3-24. To signal "hover," extend the arms horizontally to the sides, palms down. (When guiding a landing helicopter, this signal should not be given until the helicopter is at a normal hover height above the ground and just short of the desired landing point, depending on its forward speed.) (See figure 3-22.)

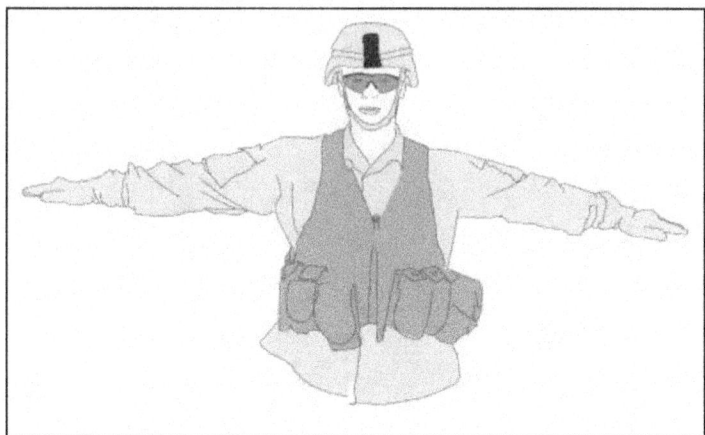

Figure 3-22. Hover

GROUND-TO-AIR PANEL SYSTEM

3-25. The panel system is a method ground troops use to communicate, to a limited degree, with aircraft by displaying panels on the ground. There are two types of panels: marking and identifying colored panels, and black and white panels for transmitting messages.

- The marking and identifying panels are made in fluorescent colors. The panels are used to mark positions and identify friendly units. These panels can be ordered through the supply system using the nomenclature "Panel Marker, Aerial, Liaison" (see figure 3-23 on page 3-13).
- Black and white panel sets. They are arranged on light or dark terrain backgrounds. They are used to transmit brief messages or to identify a unit. This is done by using the combined panel system and the panel recognition code in the unit's communications-electronics operating instructions.

3-26. Panels (if constructed locally) should be large enough to permit easy reading from the air. There should be as much color contrast as possible between the symbols and the background. Panels should be at least six feet long and two feet wide.

3-27. Select a relatively flat, clear area of ground about 40 by 130 feet. This area is large enough to display messages and special signs. For message drop and pickup, the area should be clear of obstacles which could prevent aircraft from flying into the wind at reduced airspeed and low altitude.

3-28. When using the panel system, one of the panels is used as a base panel. Place the base panels first and keep them in place as long as panel signaling is in progress. The distance between panels is one panel length throughout, when space is available. Change from one panel figure to another as soon as possible by shifting, adding, or removing panels (other than the base panels). The index panel is the first removed and the last laid out when the display is changed. Remove all panels from view that are not used for a particular display.

Hand and Arm Signals for Aircraft

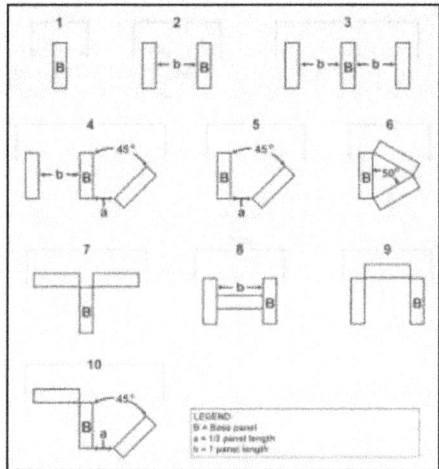

Figure 3-23. NATO standard panel code figures for numbers

3-29. The unit's electronic signal operating instructions (SOI) assign specific vocabulary, receipting, acknowledging, and identification procedures. Code meanings are normally based on these instructions, with local amplification, while the numbers associated with the meanings are determined by the unit's SOI. They are changed periodically to prevent compromise.

3-30. An aircraft pilot indicates that ground signals have been understood by rocking the wings laterally, by flashing a green signal lamp, or by any prearranged signal (A, figure 3-24). The pilot indicates that ground signals are not understood by making a 360-degree turn to the right, by flashing a red signal lamp, or by any prearranged signal (B, figure 3-24). Each panel display is acknowledged. A pilot requests a unit to display an identification code by a prearranged signal. In no case does a unit display an identification code until the aircraft has been identified as friendly.

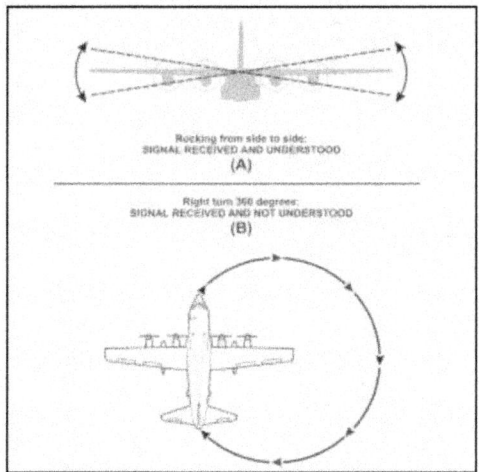

Figure 3-24. Aircraft acknowledgement

SPECIAL PANEL SIGNALS

3-31. Special panel signals are used to convey messages to rotary wing and fixed wing aircraft when the use of radio and/or digital systems are unavailable. There may be times when special panels are used for redundancy with radio and digital systems or when units are required to maintain radio silence.

WIND-T

3-32. The T is used to indicate wind direction. It represents an aircraft flying into the wind. The wind-T is two panels wide and two panels long (see figure 3-25).

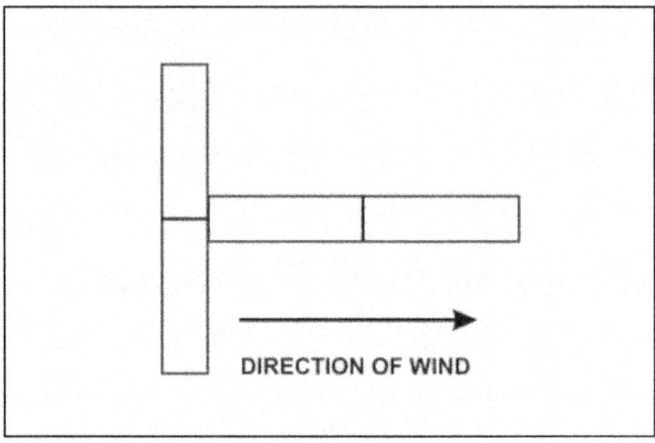

Figure 3-25. Wind direction, wind-T

MESSAGE PICKUP

3-33. This message is displayed by the figure 8 (H) with the wind-T centered below it. The crossbar of the H (8) is not placed in position until the message is ready to be picked up. The pickup poles are placed so that each pole is one panel-length away from the corner of the nearest panel (see figure 3-26 on page 3-15).

Hand and Arm Signals for Aircraft

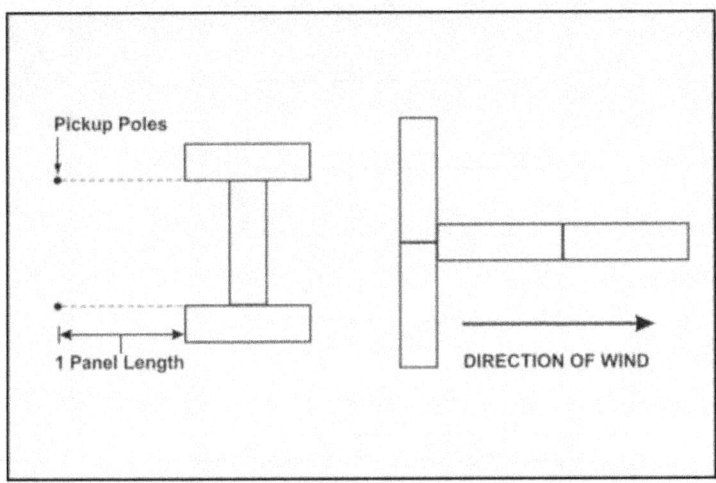

Figure 3-26. Pick up message here (wind in direction indicated)

MESSAGE DROP

3-34. When a dropped message is not found, this symbol is displayed in the drop area (see figure 3-27).

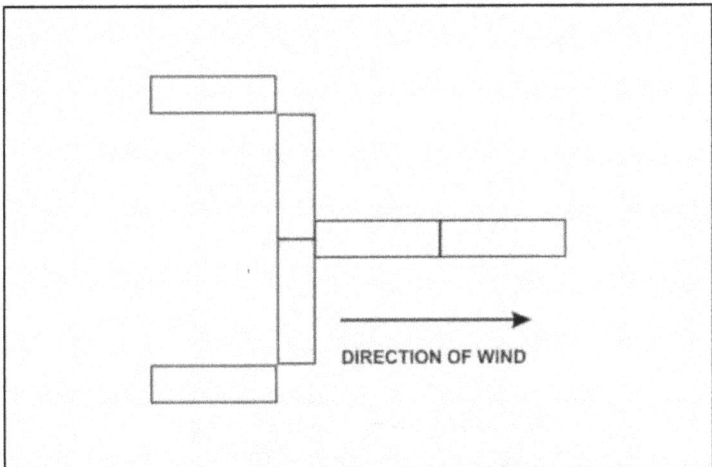

Figure 3-27. Dropped message not received

ENEMY AIRCRAFT

3-35. Two panels, placed at right angles to a third and on the axis of any base panel, always means enemy aircraft near–even though other parts of the panel display remain in place (see figure 3-28 on page 3-16).

Chapter 3

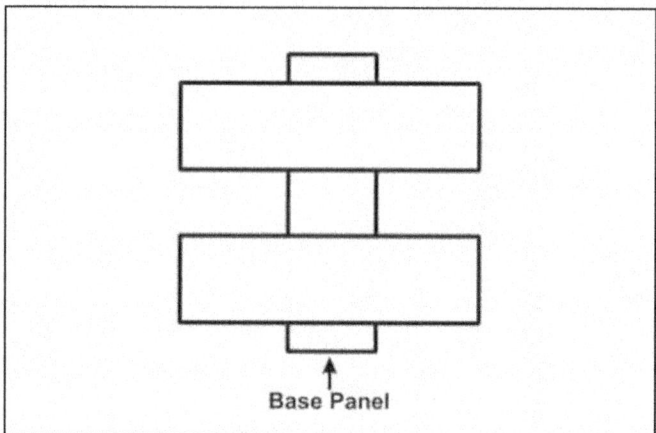

Figure 3-28. Enemy aircraft in your vicinity

DIRECTION INDICATOR

3-36. An arrow made with not less than four panels means "in this direction." This sign is used alone or with the pattern preceding it to complete its meaning (see figure 3-29).

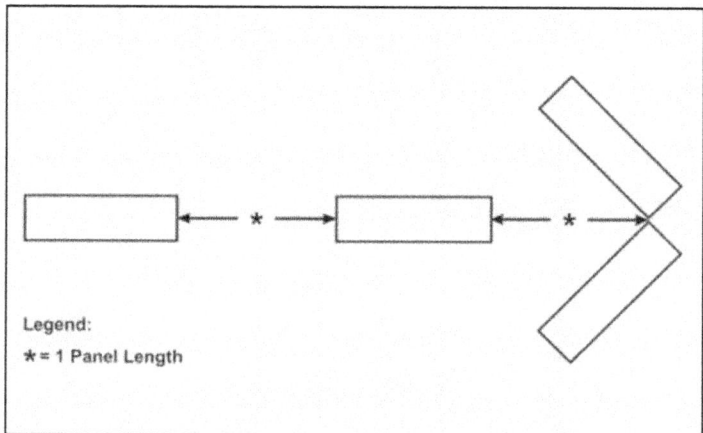

Figure 3-29. Direction indicator

GROUND-TO-AIR EMERGENCY SIGNALS AND CODES

3-37. Aviators have developed two methods of transmitting emergency messages once a pilot's attention has been obtained. These signals and codes are typically understood by all allied nations.

Hand and Arm Signals for Aircraft

EMERGENCY SIGNALS

3-38. The body can be used to transmit messages. The individual stands in an open area to make the signals. He ensures that the background (as seen from the air) is not confusing, goes through the motions slowly, and repeats each signal until it has been understood (see figure 3-30).

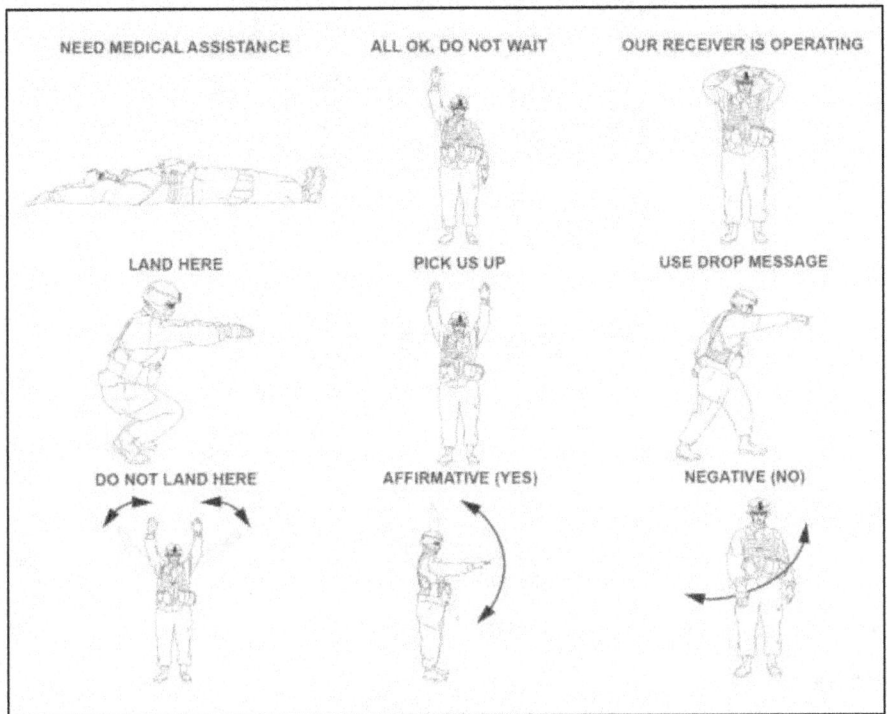

Figure 3-30. Emergency signals

EMERGENCY CODES

3-39. The symbols for these codes may be constructed from any available material that contrasts with the back-ground; for example, strips of parachute canopy, undershirts torn into wide strips, rocks, sticks, and foliage stripped from trees. Once laid out, these signals (codes) are semipermanent (see figure 3-31).

Chapter 3

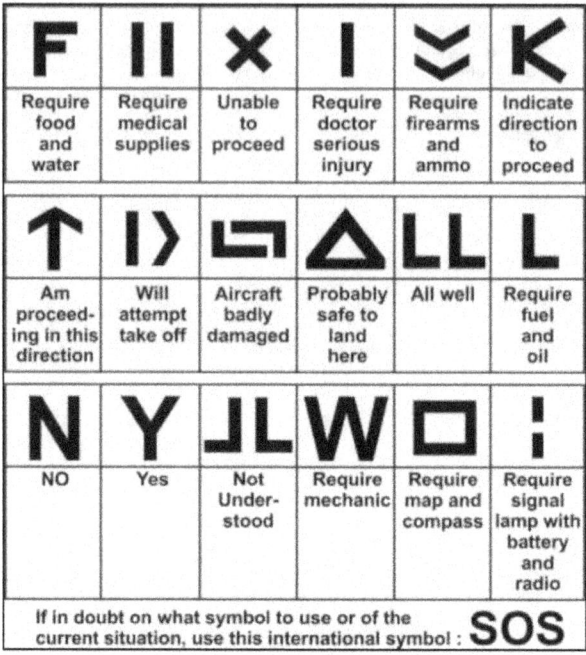

Figure 3-31. Emergency codes

SIGNALING WITH MIRRORS AND STROBES

MIRRORS

3-40. Mirrors are used to get the attention of an aircraft pilot during the day. Their use requires good visibility and little or no cloud cover in order to reflect the sun. Mirrors also can be used to transmit messages, if signals have been arranged. The MK 3 signal mirror is designed for use as a signal device. Instructions for its use are printed on the back of the mirror (see figure 3-32 on page 3-19).

Hand and Arm Signals for Aircraft

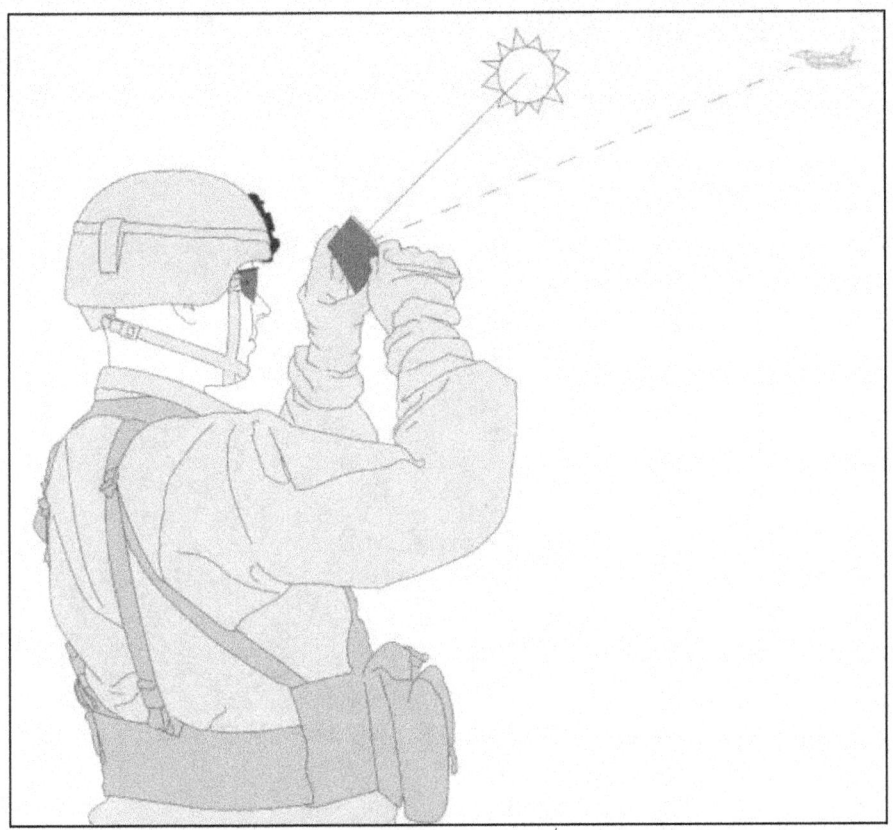

Figure 3-32. How to use a signal mirror

STROBES

3-41. Strobes can be used at night to identify positions. If prior coordination has been conducted with supporting aviation units, strobes may be used to signal pilots. To reduce detection when used, strobe lights should be placed in holes so they can only be viewed from above. Strobes with infrared covers can be used if there has been prior coordination with the aircrew. Strobes are ordered using the nomenclature, "Distress Markers."

This page intentionally left blank.

Chapter 4
Pyrotechnics

GENERAL

4-1. Pyrotechnics produce either smoke or light and are consumed in the process. When used for communications, prearranged or prescribed signals are developed and used throughout a unit. These signals are developed based on the color and characteristics of the pyrotechnic device used. Pyrotechnic signals supplement or replace normal means of communication and allow a large number of troops and/or isolated units to be signaled quickly. They can be used for friendly unit identification, maneuver element control, fire support control, target marking, and location reports. When pyrotechnics are used, the signal and its meaning are included in the command and signal portion of the operation order and in the unit's communications-electronics operating instructions.

> **WARNING**
>
> DO NOT DISCHARGE PYROTECHNICS IN THE VICINITY OF AIRCRAFT FLYING IN THE AREA.

DESCRIPTION

4-2. Pyrotechnics are usually issued as complete rounds. There are two common types of military pyrotechnics used for signaling—handheld devices and ground smoke. The 40-mm rifle mounted can fire pyrotechnic rounds. (Refer to TM 3-22.31 for more information.)

HANDHELD SIGNALS

4-3. Handheld signals are rocket-propelled, fin-stabilized, and consist of three concentric tubes. The outer tube is the container, the next is the launcher, and inside the launcher is the fin-stabilized tube containing the rocket propellant and signal element. When fired, the fin-stabilized tube is lifted about 50 feet in the air, the signal element is expelled, and it burns from 4 to 42 seconds, (depending upon the type of signal: cluster, or parachute devices).

4-4. The following types of handheld signal rockets are typically used.
- Star clusters. Star clusters are used for signaling and illuminating. They are issued in an expendable launcher that consists of a launching tube and a firing cap. These signals produce a cluster of five free-falling pyrotechnic stars. Star clusters are available in green, red, and white (see figure 4-1 on page 4-2).

Chapter 4

Figure 4-1. Star clusters

- Star parachutes. Star parachutes are used for signaling and illuminating. They are issued in an expendable launcher that consists of a launching tube and a firing cap. These signals produce a single parachute-suspended illuminant star. Star parachutes are available in green, red, and white (see figure 4-2).

Figure 4-2. Single star

- Smoke parachutes. Smoke parachutes are used for signaling only. They are issued in an expendable launcher that consists of a launching tube and a firing cap. The device is a perforated cannister that is parachute-suspended. They are available in green, yellow, and red smoke.

GROUND SMOKE

4-5. Smoke may be used for both ground and ground-to-air signaling. Both white and colored smoke may be used for this purpose. Smoke signals are visible over greater distances when employed against a terrain background of contrasting color. Smoke is valuable for marking unit flanks, positions of lead elements, and locations of targets, drop zones, tactical landing areas, and medical evacuation landing sites. Smoke signals are not suitable for messages, but are applicable when communicating by prearranged signals between small units and with aircraft. Smoke signals may be observed by the enemy; therefore, due regard for secrecy must be considered to try and avoid disclosing position locations and/or a unit's intentions.

4-6. Smoke grenades are available in white, green, yellow, red, and violet smoke. This color range is provided by two types of grenades.

- The M8 white smoke hand grenade. is a burning-type grenade used for signaling and for laying smoke screens. When ignited, it produces dense white smoke for 105 to 150 seconds. It will not normally injure exposed troops. In heavy concentrations, troops should wear the field protective mask. However, the mask will not protect against heavy concentrations of this smoke in enclosed spaces due to oxygen depletion and carbon monoxide buildup.
- The M18 colored smoke grenade is similar in appearance to the white smoke grenade, but its top is painted the color of the smoke it produces. Its filler is a burning-type mixture containing a dye; only four are standard: red, green, violet, and yellow. As a burning-type grenade, it has an igniting-type fuse, and burns for 50 to 90 seconds.

This page intentionally left blank.

Glossary

The glossary lists acronyms and terms with Army or joint definitions. Where Army and joint definitions differ, (Army) precedes the definition. Terms for which TC 3-21.60 is the proponent are marked with an asterisk. The proponent manual for other terms is listed in parentheses after the definition.

SECTION I – ACRONYMS AND ABBREVIATIONS

This section contains no entries.

SECTION II – TERMS

This section contains no entries.

This page intentionally left blank.

References

REQUIRED PUBLICATIONS

ADRP 1-02, *Terms and Military Symbols*, 16 November 2016.
AR 385-10 *The Army Safety Program*, 27 November 2013.
Department of Defense Dictionary of Military and Associated Terms, February 2017.

RELATED PUBLICATIONS

Most Army doctrinal publications and regulations are available at: http://www.apd.army.mil.

FM 27-10, *The Law of Land Warfare*, 18 July 1956.
TM 3-22.31, *40-mm Grenade Launchers*, 17 November 2010.

RECOMMENDED READINGS

Most Army doctrinal publications and regulations are available at: http://www.apd.army.mil.
Other publications are available on the Central Army Registry on the Army Training Network, https://atiam.train.army.mil.

ATP 3-20.15, *Tank Platoon*, 13 December 2012.
ATP 4-31, *Recovery and Battle Damage Assessment and Repair*, 27 August 2014.
ATP 4-35.1, *Ammunition and Explosives Handler Safety Techniques*, 8 November 2016.
FM 3-05.70, *Survival*, 17 May 2002.
FM 3-21.38, *Pathfinder Operations*, 25 April 2006.
TC 3-22.90, *Mortars*, 2017.
TC 3-23.30, *Grenades and Pyrotechnic Signals*, 22 November 2013.
TC 21-305-20, *Manual for the Wheeled Vehicle Operator* {AFMAN 24-306(I)}, 12 January 2016.
TM 4-48.09, *Multiservice Helicopter Sling Load: Basic Operations and Equipment* {MCRP 4-11.3E, Vol I: NTTP 3-04.11; AFMAN 11-223(I), Vol I; COMDTINST M13482.2B}, 23 July 2012.
TM 4-48.10, *Multiservice Helicopter Sling Load: Single-Point Load Rigging Procedures*, 5 July 2013.
TM 4-48.11, *Multiservice Helicopter Sling Load: Dual-Point Load Rigging Procedures*, 5 July 2013.
STP 7-11B1-SM-TG, *Soldier's Manual and Trainer's Guide, MOS 11B, Infantry, Skill Level 1*, 6 August 2004.
STP 21-1-SMCT, *Soldier's Manual of Common Tasks Warrior Skills Level 1*, 10 August 2015.

PRESCRIBED FORMS

This section contains no entries.

REFERENCED FORMS

Unless otherwise indicated, DA forms are available on the Army Publishing Directorate (APD) web site (www.apd.army.mil).

DA Form 2028, *Recommended Changes to Publications and Blank Forms*.

This page intentionally left blank.

Index

Entries are by paragraph number unless mentioned otherwise.

E

emergency signal
  two methods. 3-37
    emergency codes. 3-39
    emergency signals. 3-38

F

firing range
  flag signals for. 2-56
flag color
  meaning of. 2-54

H

handheld signal rocket. 5-4
  smoke parachutes. 4-4
  star clusters. 4-4
  star parachutes. 5-4

M

mechanized movement
  techniques. 2-4

P

panel system
  two types. 3-25
    black and white. 3-25
    fluorescent colors. 3-25
patrol signals. 1-18

S

signalling
  mirrors. 3-40
  strobes. 3-41
smoke grenade
  two types. 4-6
    M18 colored. 4-6
    white. 4-6

This page intentionally left blank.

TC 3-21.60
6 March 2016

By Order of the Secretary of the Army:

MARK A. MILLEY
General, United States Army
Chief of Staff

Official:

GERALD B. O'KEEFE
Administrative Assistant to the
Secretary of the Army
1705505

DISTRIBUTION:
Active Army, Army National Guard, and United States Army Reserve: Distributed in electronic media only (EMO).

PIN: 201473-000

www.ingramcontent.com/pod-product-compliance
Lightning Source LLC
Chambersburg PA
CBHW050231230526
45470CB00005B/1900